Invisibility

INVISIBILITY

THE HISTORY AND SCIENCE OF HOW NOT TO BE SEEN

• • •

GREGORY J. GBUR

Yale
UNIVERSITY PRESS
New Haven and London

Published with assistance from the Mary Cady Tew Memorial Fund.

Yale University Press books may be purchased in quantity for educational, business, or promotional use. For information, please e-mail sales.press@yale.edu (U.S. office) or sales@yaleup.co.uk (U.K. office).

Set in Adobe Garamond type by Integrated Publishing Solutions, Ltd.
Printed in the United States of America.

Library of Congress Control Number: 2022941424
ISBN 978-0-300-25042-8 (hardcover: alk. paper)

A catalogue record for this book is available from the British Library.

This paper meets the requirements of ANSI/NISO Z39.48-1992 (Permanence of Paper).

10 9 8 7 6 5 4 3 2 1

For my friend Kayla Arenas, with deepest thanks,
and my roommate Sarah Addy, in gratitude for all her support

Contents

1

In Which I Made a Bad Prediction

There is no reason for any individual to have a computer in their home.

attributed to Ken Olsen, founder of Digital Equipment Corporation, 1977

Scientific progress is somewhat notoriously hard to foresee. The history of science is filled with examples of spectacularly incorrect high-profile predictions. For example, in a series of lectures titled *Light Waves and Their Uses,* the famed physicist Albert A. Michelson said, "The more important fundamental laws and facts of physical science have all been discovered, and these are so firmly established that the possibility of their ever being supplanted in consequence of new discoveries is exceedingly remote."[1] Michelson made this statement in 1899; within a few years, however, the introduction of the special theory of relativity and of quantum mechanics would completely upend this view of physics. Albert Einstein's special theory of relativity describes how the motion of objects is very different when the relative speed of the objects is close to the speed of light, and quantum mechanics describes how the behavior of both light and matter on extremely small scales is very different from what we see in everyday life. Both theories transformed our understanding of physics and the universe, leading to new insights that to this day are still not fully understood.

Michelson cannot be seriously faulted: he was basing his statement on the available scientific knowledge of his era. But some predictions are doomed due to a simple lack of understanding of known science. An embarrassing example appeared in an editorial in the *New York Times* on January 13, 1920, in which an anonymous editor blasted Professor Robert Goddard, a pioneering researcher in rocketry, and derided his idea of sending a rocket outside the Earth's atmosphere as being impossible:

> Still, to be filled with uneasy wonder and to express it will be safe enough, for after the rocket quits our air and really starts on its longer journey, its flight would be neither accelerated nor maintained by the explosion of the charges it then might have left. To claim that it would be is to deny a fundamental law of dynamics, and only Dr. Einstein and his chosen dozen, so few and fit, are licensed to do that.
>
> That Professor Goddard, with his "chair" in Clark College and the countenancing of the Smithsonian Institution, does not know the relation of action to reaction, and of the need to have something better than a vacuum against which to react—to say that would be absurd. Of course he only seems to lack the knowledge ladled out daily in high schools.[2]

The author of this rather insulting editorial mistakenly assumed that a rocket derives its thrust from its exhaust pushing off of the atmosphere. A rocket in space, however, derives its thrust from Newton's Third Law of Motion, which states that every action has an equal and opposite reaction: the backward-moving exhaust of the rocket results in the rocket moving forward. On July 17, 1969, as the Apollo astronauts headed to their historic first landing on the Moon, the *Times* issued the following correction: "Further investigation and experimentation have confirmed the findings of Isaac Newton in the 17th century and it is now definitely established that a rocket can function in a vacuum as well as in an atmosphere. The Times regrets the error." So predictions about science are doubly perilous: they can turn out to be wrong because of science that is currently unknown, and they can turn out to be wrong because of a fundamental misunderstanding of known science.

With this in mind, I should have known better myself when I was asked to comment on future science.

On May 25, 2006, two groups of researchers announced independent papers in the journal *Science* outlining theoretically how one could design an invisibility cloak. The first paper, "Optical Conformal Mapping," was written by Ulf Leonhardt, then at the University of Saint Andrews in Scotland; the second, "Controlling Electromagnetic Fields," was written by John Pendry of Imperial College, London, together with David Schurig and David Smith of Duke University in North Carolina. Despite the rather technical-sounding titles, the papers had exciting implications: both described similar strategies for how one could design a device that would guide light around a central hidden region and send it along on its way, as if it had encountered nothing at all. The device would in principle "cloak" an object within it from detection, making it effectively invisible.[3]

I was perhaps more excited than anyone else about these results, since my own PhD work, completed in 2001, was related to early attempts to describe invisibility in physics. In 2003, in a meeting in Kiev, Ukraine, I happened to talk a little about my early work with Ulf Leonhardt, a fellow attendee. The 2006 research, however, was a game changer and really made scientists consider that invisibility might not only be possible but actually be feasible.

Naturally, the two publications captured worldwide attention, and journalists and scientists everywhere scrambled to understand the implications of the results. Thanks to my earlier work in the field (which we'll talk about later), I was contacted by a few news organizations to discuss invisibility cloaks and their potential uses. It was during one of these interviews that I was asked the perilous question: "When do you think we will see a working invisibility cloak?"

Facing such a question, the scientist has a natural instinct to be conservative. The technology required to make invisibility cloaks a reality did not seem to exist yet in 2006 and appeared to be quite difficult to achieve in practice. Any answer I gave would be at best a guess, and I settled on "five years" for my reply. This number reflected my view that the experiments would be difficult and take time but would not be impossible. And if I were wrong and experiments were never successful, nobody would remember what I had said after five years.

The first experimental demonstration, however, was published in November 2006, only six months after the theoretical work was released—I had guessed too long by four and a half years![4] Though that first experimental test was done using microwaves, not visible light, and was thus not invisible in the strictest sense of the word, it showed that the principles of invisibility cloaks are not as impossible as they first seemed. This, for me, exemplifies a trend in the science of invisibility ever since: it has been full of surprises and has been hard to predict.

Since those groundbreaking papers in 2006, scientific journals and news sources alike have been filled with breathless accounts suggesting that true invisibility is almost here, perhaps just one crucial discovery away from being a reality. A small sample of some of my favorite headlines:

- "Invisibility Cloaks Are in Sight" (May 25, 2006)
- "Researchers Create Functional Invisibility Cloak Using 'Mirage Effect'" (October 5, 2011)

- "We're One Step Closer to a True Invisibility Cloak" (January 26, 2012)
- "Scientists Invent Harry Potter's Invisibility Cloak—Sort Of" (March 30, 2013)
- "Canadian Company Created an Invisibility Shield Called 'Quantum Stealth'" (October 21, 2019)

Special distinction goes to my favorite headline of invisibility physics: "'Invisibility Cloak' Makes Tanks Look Like Cows" (June 9, 2011).

Considering such lurid titles, you might not unreasonably begin to wonder whether someone is standing behind you right now, reading over your shoulder (don't worry—they aren't). What is the current state of research into invisibility, and how does it work? Or does it work at all? This book tackles these and other major questions.

And there is more to the story: scientists and science fiction authors have studied the phenomenon of invisibility and tried to understand how it might work for more than 150 years. We will trace this history, following the same steps that scientists themselves took to understand the nature of both light and matter. Along the way, we will see how science fiction writers anticipated some of the most remarkable discoveries in the field. In the end, we will find that the science of invisibility is much stranger and more unexpected than even the most visionary science fiction authors imagined.

2
What Do We Mean by "Invisible"?

As she said this, she looked up, and there was the Cat again, sitting on a branch of a tree.

"Did you say pig, or fig?" said the Cat.

"I said pig," replied Alice; "and I wish you wouldn't keep appearing and vanishing so suddenly: you make one quite giddy."

"All right," said the Cat; and this time it vanished quite slowly, beginning with the end of the tail, and ending with the grin, which remained some time after the rest of it had gone.

"Well! I've often seen a cat without a grin," thought Alice; "but a grin without a cat! It's the most curious thing I ever saw in my life!"

Lewis Carroll, Alice's Adventures in Wonderland *(1865)*

The first scientific papers describing the physics of invisibility cloaks were published in 2006, and they were widely and correctly regarded as revolutionary for physics. So one can imagine my genuine shock when, while researching this book, I came across an article titled "Cloaks of Invisibility," published in 1944 in the *Science News-Letter.*[1]

Despite the provocative title, this paper concerned something much more mundane: the Allies' use of clever camouflage techniques during World War II. An illustration of a "spider-hole" was captioned: "If you were a soldier in the peaceful field shown in the top picture

Figure 1. A "spider-hole." Illustration from Thone, "Cloaks of Invisibility" (1944).

and saw no sign of the enemy, you might be very close indeed to death" (fig. 1). Such holes are named for the burrows created by a variety of spiders, such as those of the family Ctenizidae. The arachnid burrows possess a camouflaged cover and a hinge of silk that allows the spiders to easily ambush prey, very much the way that soldiers use their holes to ambush enemy troops.

Once I got over my shock, I realized that my discovery highlighted a problem with any scientific discussion of invisibility: the word "invisible" is simultaneously very suggestive, conjuring a specific image (or lack thereof) in a person's mind, and very vague, in that it can mean many different things. The word "invisible" means "unable to be seen," but there are many ways this can be accomplished without sophisticated physics. If you hide in a spider hole, you are invisible; if you shut off the lights in a windowless room, you are invisible. In the classic Monty Python sketch "How Not to Be Seen," subjects achieve invisibility by hiding behind bushes. Very small objects, such as bacteria, atoms, and molecules, cannot be seen with the naked eye; they, too, are invisible. Animals such as octopuses and chameleons achieve a sort of invisibility by matching their skin coloration and patterns to the background. Cover your eyes? Everything is invisible to you.

Clearly, we will want to restrict our usage of the term, and here an example may help us with this effort. In 2003, before the famous cloaking papers were published, another "coat of invisibility," created by Professor Susumu Tachi of the University of Tokyo, made world headlines (fig. 2). The effect of the coat is quite eerie: a user wearing this cloak of invisibility appears to be partly transparent and can move freely, maintaining the illusion regardless of their body's position.[2]

The invisibility effect is created with a relatively simple set of tools. The coat is made of a retroreflective material that directs most of the light hitting it back toward its source; this is combined with a camera recording the scene behind the coat wearer and transmitting it to a projector in front of the wearer. The projector displays the background scene on the coat, giving the illusion of transparency, in what the paper dubbed "retroreflective projection technology."

Figure 2. A demonstration of the "optical camouflage" system. Photographs courtesy of Susumu Tachi, the University of Tokyo.

In this form, the invisible coat is a far cry from the sort of technology that might be used in spying or warfare: the illusion is effective only when viewed from the position of the projector itself and will look disjointed from other positions. But sinister applications were far from Tachi's mind in his research: "Asked if this invisible technology could have military applications, say in a desert war, the 57-year-old professor flinched, quite visibly. Reflecting Japan's deeply held pacifist ethic, universities generally shun military research."[3]

Instead, Tachi's coat grew out of his research on telexistence, the use of technology to connect and control objects in remote environments, or use a virtual environment to augment a real environment. Foremost on his mind was the use of the technology to enhance surgery: if retroreflective projection were combined with magnetic resonance imaging (MRI) of the human body, the internal organs of a patient could be projected right on their skin, making it easier for a surgeon to make the proper incisions.

Another intriguing possibility advanced and tested by Tachi is the idea of making the interior of a car or airplane cockpit transparent, so that the pilot of the craft can have full vision in all directions and, in the case of cars, a precise awareness of the location of nearby obstacles. More recently, a similar idea was introduced and prototyped by fourteen-year-old Alaina Gassler, who designed a projection system to make the A-pillars of a car transparent; these pillars normally create blind spots in the driver's line of sight. Gassler presented her results at the Broadcom MASTERS science and engineering middle school competition and deservedly took home the $25,000 top prize.[4]

This type of invisibility is what we will term "active invisibility," in which the device measures the light illuminating the object to be concealed and then generates new light to create the illusion. This is in contrast to much of what we will talk about throughout the book—"passive invisibility," in which the device only manipulates and guides the illuminating light. Such active invisibility schemes like Tachi's and Gassler's have been used for more frivolous purposes: in 2012, Mercedes-Benz created an "invisible car" to promote its low emission hydrogen-powered vehicles.[5] This car had a camera on the right side to record the scene and an array of LEDs on its left side to project the recorded image. As in the earlier examples, this illusion worked only when the car was viewed from the proper angle—in this case, from directly on its left side—and it was in most circumstances far from being truly unseen. The idea here may have been inspired by the Aston Martin V12 "Vanish," James Bond's invisible car that appeared in the movie *Die Another Day* (2002).

But this brings us back to the question of what we mean by invisible. Neither Tachi's coat nor Mercedes-Benz's car are truly invisible in the strict sense of the word. Both are more see-through (transparent), and their illusions are effective only when observed from a particular direction. But they do represent efforts to use science and technology to make an object harder to see, and it would be a shame to exclude them from our definition; furthermore, as we will find, it is not entirely clear that it is possible, even in principle, to make an object perfectly invisible. Many of the types of invisibility we will discuss throughout the book have significant limitations. For example, the object may appear perfectly invisible only when viewed or illuminated from a certain angle, like Tachi's coat, or it may be invisible only for a certain range of colors of light.

So for our working definition, let us try the following: we will consider objects invisible if they manipulate light in an unusual way to make an object harder to see than we would normally expect. This excludes simple cases like hiding behind a sofa or switching off a light but includes intriguing examples such as those discussed above and others to come.

There is one other important lesson to take away from Tachi's invisible coat. From science fiction stories, we are used to thinking of invisibility as a sinister power, one that can be used to cause great harm. But invisibility also has a surprising number of beneficial applications, such as reducing car accidents or aiding in surgery. We will see more examples of these unexpected uses to come.

3
Science Meets Fiction

Little did he dream what was about to happen when he entered an arbour, thickly covered with jessamine, woodbine, and roses, and threw himself upon a seat, where he had passed many trance-like hours with Alicia. "I should like to know what they are now talking about," he exclaimed. "I wish I was invisible!"

No wish could possibly be much more ridiculous, but it struck his fancy at the moment, and he again repeated it; and then allowing his imagination to play with the idea for a minute or two, he became highly excited by the sport which it presented to his view, and again he ejaculated, "What a glorious thing it would be! I do, indeed, wish I could be invisible!"

The number three has long been celebrated for its potency, both for good and evil; and no sooner had the third exclamation passed his lips, than he heard a short cough, not many yards from the place where he was sitting. Instantly starting up, he looked out from among the clustering tendrils, and beheld a stranger, walking slowly towards the bower.

James Dalton, The Invisible Gentleman *(1833)*

In the classic movie *Clash of the Titans* (1981), the young man Perseus awakens unexpectedly in the amphitheater of the city of Joppa, cast there by the jealous goddess Thetis. Perseus's father is Zeus, the ruler of all the gods, and he is not happy with Thetis's meddling. To help protect his son now that he is out in the greater world with all its dangers, Zeus sends three gifts down to the amphitheater for Perseus: a sturdy mirrored shield, a sword strong enough to cleave marble, and a helmet that renders its wearer invisible. By the end of his adventure, Perseus will find good uses for all three items.

The story of *Clash of the Titans* is loosely based on ancient Greek folklore, dating back two thousand years. The legend of Perseus evolved over centuries, but one of the most influential versions was written in the first or second century CE in the *Bibliotheca* of an author now known as Pseudo-Apollodorus. (The author was originally thought by scholars to be Apollodorus of Athens, but since then evidence has proven this not to be the case. So the unknown author is known as "false" Apollodorus.) In the *Bibliotheca,* Perseus uses the cap of Hades, which grants invisibility, in order to sneak up on Medusa and her sisters and escape with her head:

> Wearing it, he saw whom he pleased, but was not seen by others. And having received also from Hermes an adamantine sickle he flew to the ocean and caught the Gorgons asleep. They were Stheno, Euryale, and Medusa. Now Medusa alone was mortal; for that reason Perseus was sent to fetch her head. But the Gorgons had heads twined about with the scales of dragons, and great tusks like swine's, and brazen hands, and golden wings, by which they flew; and they turned to stone such as beheld them. So Perseus stood over them as they slept, and while Athena guided his hand and he looked with averted gaze on a brazen shield, in which he beheld the image of the Gorgon, he beheaded her. . . . So Perseus put the head of Medusa in the wallet (*kibisis*) and went back again; but the Gorgons started up from their slumber and pursued Perseus: but they could not see him on account of the cap, for he was hidden by it.[1]

People have therefore been imagining the benefits—and drawbacks—of having the power of invisibility for millennia, if not longer.

A darker take on invisibility appears in Plato's grand philosophical work *The Republic,* written about 375 BCE. In a dialogue with Socrates, the character Glaucon presents the story of the ring of Gyges, a story that on the surface will sound suspiciously familiar to fans of fantasy fiction:

> According to the tradition, Gyges was a shepherd in the service of the king of Lydia, and, while he was in the field, there was a storm and earthquake which made an opening in the earth at the place where he was feeding his flock. Amazed at the sight, he descended into the opening, where, among other marvels, he beheld a hollow brazen horse, having doors, at which he stooping and looking in saw a dead body of stature, as appeared to him, more than human, and having nothing on but a gold ring; this he took from the finger of the dead and reascended. Now the shepherds met together, according to custom, that they might send their monthly report about the flocks to the king; into their assembly he came having the ring on his finger, and as he was sitting among them he chanced to turn the collet of the ring inside his hand, when instantly he became invisible, and the others began to speak of him as if he were no longer there. He was astonished at this, and again touching the ring he turned the collet outwards and reappeared; thereupon he made trials of the ring, and always with the same result; when he turned the collet inwards he became invisible, when outwards he reappeared. Perceiving this, he immediately contrived to be chosen one of the messengers sent to the court, where he no sooner arrived than he seduced the queen, and with her help conspired against the king and slew him, and took the kingdom. Suppose now that there were two such magic rings, and the just put on one of them and the unjust the other; no man, they say, is of such an iron nature that he would stand fast in justice.[2]

Plato, through Glaucon and the story of Gyges, wonders whether virtue exists only due to the fear of being punished and that justice itself is therefore only a construct of society. Plato, through Socrates, answers this by arguing that the man who succumbed to the temptation of ultimate power would have effectively punished himself by being enslaved to his own base desires. The man who chooses not to succumb keeps control of himself and is therefore happy and free.

These stories of the power of invisibility, and the corrupting nature of that power, have been mirrored numerous times since these ancient tellings. Plato, Pseudo-Apollodorus, and those who followed viewed invisibility as a gift from the gods, the result of magic, an imaginary conceit that could be used as metaphor or example but not reality. In *The Invisible Spy,* written pseudonymously by Eliza Haywood in 1754, the narrator acquires an invisible belt from a mystical magus; in *The Invisible Gentleman,* written by James Dalton in 1833, the protagonist gains invisibility through a rash wish.

As humanity's understanding of the natural world increased, it was inevitable that someone would try to imagine whether invisibility was possible within the bounds of natural law. It was not a scientist who first asked this question, however, but an author of science fiction. In 1859, Irish-born American author Fitz James O'Brien published "What Was It? A Mystery," which has the distinction of being the earliest story ever written that attempts to give a scientific explanation for invisibility (fig. 3).

The life of Fitz James O'Brien was wild, turbulent, and unpredictable, and it was reflected in the variety of writing he did and its occasional strangeness. He was born in Ireland in 1828 as Michael O'Brien, the son of an attorney-at-law. He showed an early passion for writing poetry, which would serve him well in years to come. He attended the University of Dublin and moved afterward to London, where he apparently lived a life of luxury, depleting the inheritance his father left him in only two years. This led him to seek his fortune in the United States around 1852, changing his first name to Fitz James

Figure 3. Fitz James O'Brien, by William Winter. Illustration from Winter, *Poems and Stories of O'Brien* (1881).

in the process. Fortunately, he had enough influence at home to obtain letters of introduction into the literary world of New York City, and soon he was writing for a variety of publications, including the *Evening Post* and the *New York Times,* as well as *Vanity Fair* and the *Atlantic Monthly.*

A literary career was inadequate to sustain the luxurious lifestyle that O'Brien had become accustomed to, and reports suggest that he was in debt much of the time, lived wherever he could find someone to give him room, and generally had more hard times than good. He was also a man of great charisma but poor temper, making lifelong friends with some people he encountered and enraging others. An excellent anecdote of his dual nature was provided by his friend Thomas Davis years after O'Brien's death:

> Donald McLeod, author of "Pynnshurst," was once O'Brien's comrade, and they slept in the same bed. One night, just after they had retired, a fierce discussion arose between them with reference to Scotch and Irish nationality, and O'Brien uttered opinions which

> his Scotch companion could not brook. "I'll not allow this," cried McLeod. "Do as you please about that," said O'Brien. "I'll demand satisfaction, sir!" roared McLeod. "Very well," answered Fitz-James,—equally enraged and belligerent, and pulling the blanket well over himself,—"very well, sir; you know where to find me in the morning." This last explosion, though intended in deadly sincerity, had the effect of turning the quarrel to laughter, and so made an end of it.[3]

When the Civil War broke out in 1861, O'Brien joined the Seventh Regiment of the National Guard of New York, hoping to be sent to the front lines to fight for the Union. The regiment was only deployed to guard Washington, however, and was sent home after a month of duty. Undeterred, O'Brien sought an appointment to the staff of a general and managed to secure a position with General Frederick W. Lander on the front lines in Virginia at the start of 1862. There, he joined the general in a charge against Confederate forces at Bloomery Gap, in present-day West Virginia, on February 14, being one of only a handful who were brave enough to ride into enemy fire. Several days later, O'Brien was leading a cavalry company to seize cattle belonging to the enemy when his group came upon a hostile force that outnumbered them more than four to one. Undaunted, O'Brien led a charge. "As they rode forward, the rebel officer held up his hand and cried, 'Halt! who are you?' O'Brien shouted back in reply, 'Union soldiers!' and fired at him. This was the signal for a general engagement. The rebels could easily have captured so small a party; but O'Brien's onslaught was so audacious that they thought he must have reserves somewhere."[4] The Union forces managed to drive off the enemy, but O'Brien himself was not so lucky. He killed the leader of the rebel forces and was shot in the shoulder in return. The wound was serious, but at first it appeared that Fitz James would recover. His condition gradually worsened, however, and he died on April 6, 1862, only thirty-five years old.

The strange and imaginative stories O'Brien left behind hint at a career cut short before it reached its prime. His ghost story "The Lost Room" (1858) describes a man who encounters malevolent spirits occupying the study of his home; when he rashly makes a bet with the spirits for possession of the room, he loses the chamber entirely from the physical plane. This story feels like a metaphor for O'Brien's life of wandering and loss. In "From Hand to Mouth," a struggling writer seeks shelter from a winter storm in a hotel that seems to be alive and watching him; this story, a surreal satire, is also semiautobiographical as it looks cynically at the plight of writers struggling to make a living. In "The Wondersmith," a group of evil sorcerers conspires to bring toys to life to commit murder, only to find their puppets turn against them.[5]

O'Brien's most influential stories, however, are those that involve scientific ideas, and optics in particular. His most famous work, "The Diamond Lens" (1857), tells the tale of a man who builds the most powerful microscope ever conceived and falls in love with the microscopic woman he sees in a drop of water.[6] But "What Was It? A Mystery," published in *Harper's Magazine,* is the story that truly broke new ground in science fiction in its attempt to explain the traditionally unexplainable power of invisibility.[7]

In the story, the narrator Harry and his friend Hammond take rooms in a boardinghouse that is reported to have become haunted in recent years. One night in bed, Harry is set upon in the darkness by a creature of short stature but incredible strength. He grapples with the humanoid and manages to overpower it and pin it to the floor. With a free hand, he turns on the light, only to find that he has been fighting with a living, breathing assailant that is truly invisible!

Hammond arrives and the two of them manage to tie up the creature securely and place it on the bed. They are unsure what to do with their captive, and the creature does not attempt to eat any of the food offered to it. It gradually becomes weaker and dies, and the men bury it in the backyard of the house.

As a horror story, it is not particularly scary. But as groundbreaking science fiction, it is much more compelling. After they securely bind the creature, Hammond tries to calm his friend Harry by speculating on how such an invisible being can exist:

> Let us reason a little, Harry. Here is a solid body which we touch, but which we cannot see. The fact is so unusual that it strikes us with terror. Is there no parallel, though, for such a phenomenon? Take a piece of pure glass. It is tangible and transparent. A certain chemical coarseness is all that prevents its being so entirely transparent as to be totally invisible. It is not *theoretically impossible,* mind you, to make a glass which shall not reflect a single ray of light,—a glass so pure and homogeneous in its atoms that the rays from the sun shall pass through it as they do through the air, refracted but not reflected. We do not see the air, and yet we feel it.

In his attempted explanation of invisibility, O'Brien references the two oldest experimentally observed phenomena in optics: the law of reflection and the law of refraction, both of which were discussed in the same era in which Pseudo-Apollodorus was waxing poetically about the achievements of Perseus. The law of reflection states that the angle at which a ray of light reflects from a flat, smooth surface (such as polished metal or glass) is equal to the angle at which the ray was incident upon the surface. This law, which was probably recognized in ancient times, appears in the book *Catoptrics,* long thought to have been written by the famed geometer Euclid around 300 BCE.

However, it has been argued that this book may have in fact been partly or fully written by an author or multiple authors centuries later, so the author is now referred to as Pseudo-Euclid.

The law of refraction describes quantitatively how the direction of light changes when it passes from one transparent medium to another. The phenomenon of refraction was recognized in antiquity; in *The Republic,* Plato remarks how "the same objects appear straight when looked at out of the water, and crooked when in the water."[8] Today we refer to this as the bent straw illusion, in which a drinking straw appears to have a kink in it at the air-water interface in a glass of water (fig. 4). The light that reflects from the straw underwater changes direction as it leaves the water, making the straw appear to be bent. Many scholars worked to quantify the law of refraction over the centuries; today credit is generally given to the Dutch astronomer Willebrord Snellius, who derived the correct formula in 1621. However, this law has been discovered and rediscovered many times, and the Islamic scholar Ibn Sahl is recognized as being the first to publish it some six hundred years earlier, in 984. Nevertheless, the law is often referred to as Snell's law.[9]

We will not concern ourselves with the detailed math here, but we can say that when a ray of light goes from a rarer medium (where the speed of light is higher) to a denser medium (where the speed of light is lower), its direction bends toward the line perpendicular to the surface.[10] The opposite happens when going from a denser medium to a rarer medium, and the light ray bends away from the perpendicular. The amount by which the speed of light is reduced from its value in vacuum is called the index of refraction, and it has a value of 1.33 for water, 1.5 for glass, and 2.417 for diamond. A higher refractive index results in a greater bending of the ray of light. Diamond has the highest refractive index of materials found in nature; the strong refractive power of diamond is why O'Brien envisioned a diamond

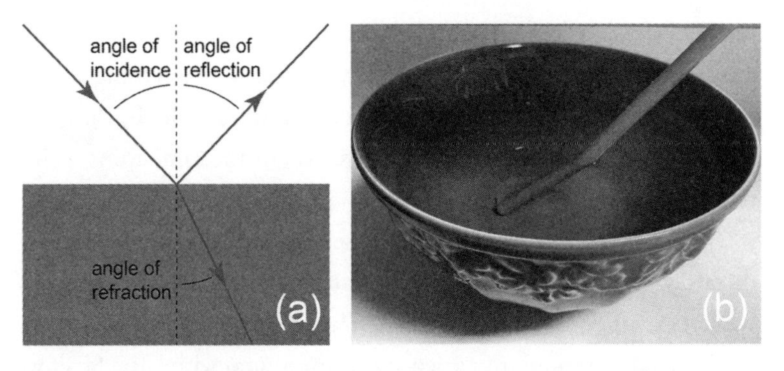

Figure 4. (a) The angles of reflection and refraction; (b) the "bent straw" illusion due to refraction.

lens producing a microscope of unprecedented power (though it turns out that making a powerful microscope is much more complicated than the choice of lens material). The index of refraction of air is around 1.0003, making the speed of light in air only slightly lower than that in vacuum; we will see later that this is a major difficulty in constructing invisibility devices.

It is worth noting that reflection and refraction go hand in hand: when light reflects off of a smooth, transparent surface, some fraction of the light is reflected and the remainder is refracted.

O'Brien's concept of invisibility is based on two key assumptions. The first of these is that the light that gets refracted through an object is otherwise unaffected by passing through the object. O'Brien may very well have been looking at a windowpane as he pondered his story; it is true that rays of light passing through a window produce very little distortion of the scene behind it. However, if light passes through a transparent curved object, such as a lens or an invisible monster, the situation is quite different! On the way out of the substance, the rays of light hit a surface inclined at a different angle than the one they entered, and so they leave the object at an angle different than the one they entered. This change of direction is what makes a lens focus light, and it would distort the image of anything standing behind a transparent monster, making it detectable. In this, Fitz James O'Brien was wrong: transparency is not the same thing as invisibility.

This problem with invisibility based on refraction would be addressed by another author decades later. Jack London, known for wilderness stories such as *The Call of the Wild* (1903) and "To Build a Fire" (1908), ventured into science fiction in 1906 with "The Shadow and the Flash," about two rival scientists who come up with distinct versions of invisibility, each with its own limitations (fig. 5).[11] Lloyd Inwood coats himself with a perfectly black paint, meaning that no light will be reflected from him, but he cannot avoid the shadow that his body casts. Paul Tichlorne, in contrast, devises a chemical process by which he can make his body perfectly transparent—but the light that passes through his body produces bright flashes as he moves, by refraction: "'They're a large family,' he said, 'these sun dogs, wind dogs, rainbows, halos, and parhelia. They are produced by refraction of light from mineral and ice crystals, from mist, rain, spray, and no end of things; and I am afraid they are the penalty I must pay for transparency. I escaped Lloyd's shadow only to fetch up against the rainbow flash.'" By the end of the story, the two rivals come to fatal blows on a tennis court, producing a bizarre spectacle of light versus darkness.

Incidentally, London's idea of a perfectly absorbing paint turned out to be prophetic. In 2014, British engineering firm Surrey NanoSystems introduced a new material dubbed Vantablack, which at the time became the record holder for the world's blackest material, absorbing 99.965 percent of the light illuminating it.[12] The material is made of a "forest" of densely packed, vertically aligned hollow carbon tubes around a billionth of a meter in diameter (a nanometer); the

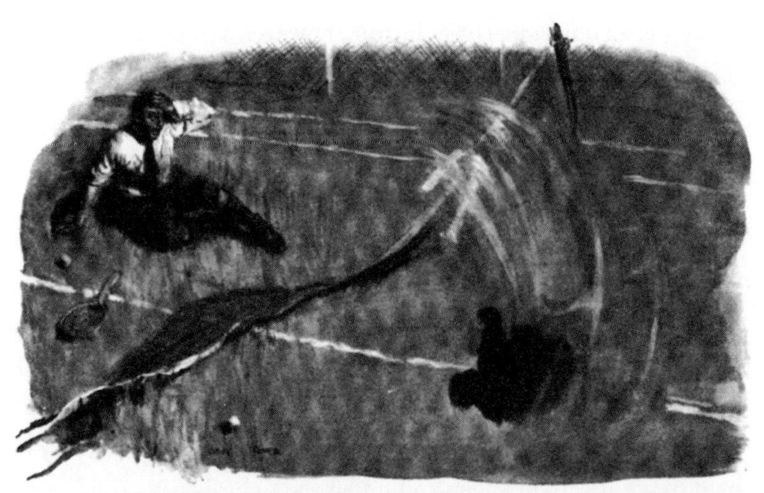

"I could do nothing, so I sat up, fascinated and powerless, and watched the struggle."

Figure 5. Original illustration from Jack London's story "The Shadow and the Flash," drawn by Cyrus Cuneo.

name Vanta is an acronym for "vertically aligned nanotube arrays." When light enters this forest, it effectively becomes lost, bouncing around multiple times between tubes before getting absorbed.

Vantablack almost immediately caused a controversy when artist Anish Kapoor procured the exclusive rights to use a spray-on version of the material in art, naturally enraging other artists. Fortunately for the art world, alternative ultrablack paints soon became available, some even blacker than Vantablack; in 2019, MIT engineers came up with a carbon nanotube–based material that absorbs 99.995 percent of incoming light, a new record for blackness. It was immediately used by artist Diemut Strebe for an exhibit at the New York Stock Exchange called "The Redemption of Vanity," in which a $2 million diamond was coated to appear as an impenetrable black void—the shadow beating the flash.[13]

One might wonder at the plausibility of even a superblack paint rendering an object invisible. In 2018, however, a museum patron fell into an exhibit by Anish Kapoor entitled *Descent into Limbo,* evidently not able to recognize that the black spot on the floor was actually a hole that was 2.5 meters (8 feet) deep.[14] And, notably, this work used not Vantablack, but a highly black paint of an earlier design.

The second key assumption that O'Brien makes about invisibility is that it is possible to make a glass so perfect that light is "refracted but not reflected." Here again, modern optics shows that O'Brien is wrong: an ordinary material like glass, no matter how pure, will always reflect at least a little bit of light (we will come to non-ordinary materials later on). Even the cleanest glass window, though it may be hard to see at certain angles and under the right lighting conditions, will reflect enough light to be detectable from some angle—it is far from invisible.

O'Brien can be forgiven for his mistakes; he was working with the limited information available to him in the era about light and its interaction with matter. But his statement about ordinary glass having a "certain chemical coarseness" that makes it visible suggests that he had some very specific physics in mind when he invented his invisibility theory, and one is naturally curious about the source of his ideas.

Unfortunately, we cannot reference O'Brien's own correspondence for a clue to his thinking. On his deathbed, he appointed two executors of his literary estate, and after his death, all of his existing writing was sent to one of those executors, Frank Wood, to be read through. Not long afterward, however, Frank Wood himself suffered an untimely death, and the collection of documents he held was evidently lost.[15]

There was so little physical understanding of the interaction of light and matter in O'Brien's day, though, that we can quickly trace his argument back to the father of modern physics himself, Isaac

Newton (1642–1727). Newton, an English mathematician and physicist, earned that title with his most famous work, *Philosophiae Naturalis Principia Mathematica* (1687), in which he introduced for the first time a universal law of gravitation and showed mathematically how it could explain the motion of not only such celestial objects as planets and moons but also objects falling near the surface of the Earth. This idea, of unifying a diverse range of observations under a small number of physical laws, became a guiding principle for all of physics through to modern times. Today, particle physicists are striving to demonstrate that the four fundamental forces of nature—gravity, electromagnetism, and the weak and strong nuclear forces—are all manifestations of a single force, in what is called a unified field theory. This effort is a direct reflection of Newton's groundbreaking efforts.

Newton was not only an excellent mathematician and theoretical physicist: he also did detailed experimental work in optics. Much of the research on optics before Newton's time focused on the geometric properties of light and studying how curved lenses and mirrors can form images of objects through reflection and refraction. Newton's optics work was pioneering in that he concentrated on understanding the physics of light, essentially trying to answer the question, "What *is* light?" His most famous contribution is the recognition that white light consists of a combination of all the visible colors of light together, which he demonstrated by splitting a beam of white light into a rainbow using a glass prism. Anyone who has seen the cover of Pink Floyd's classic album *Dark Side of the Moon* has seen an illustration of this in action.[16] The visible light spectrum contains the colors red, orange, yellow, green, blue, and violet, and the index of refraction is lowest for light on the red end of the spectrum and highest on the violet end.[17] Thus, the different colors within white light get refracted in different directions on passing through a prism.

Newton reported the first of his experiments in 1672 at the Royal Society in London, where they were immediately met with resistance. At that time, optics researchers were furiously divided over whether light should be viewed as a flow of particles, like the flow of water in a stream, or a continuous wave, acting like ripples on the surface of a pond. The particle theory was bolstered by the fact that light travels in straight lines when far from matter, while the wave theory was bolstered by its ability to explain refraction as resulting from the change of wave speed as it passes from one medium to another. Newton argued that his results demonstrated that light is a flow of particles.

Many prominent members of the Royal Society were firmly of the belief that light is a wave, and they attacked Newton's work. One of the fiercest of his critics was Robert Hooke (1635–1703), a brilliant researcher who contributed to numerous fields, including microscopy, astronomy, paleontology, mechanics, and horology (the study of time measurement). His book *Micrographia,* published in 1665, contains the earliest illustrations of microscopic organisms, and it secured his reputation as a scientist.

Hooke's fierce criticism of Newton led to a decades-long rivalry that included not only scientific criticism but accusations of plagiarism when Hooke accused Newton of stealing some of his ideas on optics and gravitation. The stress of this feud, along with other events, led Newton to a nervous breakdown in 1678, and he retired for a number of years from public life. The recognition of his work on gravitation, however, gave him celebrity status, and revived his career. The rivalry between Hooke and Newton even continued after death: Newton became president of the Royal Society after Hooke died, and it is suspected that Newton is responsible for the removal or destruction of the only known portrait of Hooke, which had been hanging at the Royal Society.

Newton finally published his work *Opticks* in 1704, not coincidentally the year after his chief rival, Hooke, had died. Among the many experiments presented in *Opticks* are attempts to understand the nature of transparency: Why are some objects see-through and others not?

In the work most relevant to our discussion, Newton explored the transparency of the smallest constituents of matter he could find. Today we would think of atoms, but in Newton's case, he speaks vaguely of the "least parts" of matter. He took transparent materials, like glass, and ground them into powder, observing that they then became opaque. In reverse, he took opaque materials, like paper, and soaked them in oil, observing that they then became transparent (if you've ever eaten greasy pizza on a paper plate, you've experienced this latter case).

Newton concluded that the least parts of matter are generally transparent but that multiple reflections and refractions of light in those least parts prevent light from getting through. A powdered glass is opaque because of the reflections and refractions that occur as light passes between particles; a slab of glass is transparent because the fused material is missing all those boundaries between particles.

In Newton's own words,

> But farther, that this discontinuity of parts is the principal Cause of the opacity of Bodies, will appear by considering, that opake Substances become transparent by filling their Pores with any Substance of equal or almost equal density with their parts. Thus Paper dipped in Water or Oil, the *Oculus Mundi* Stone steep'd in Water, Linnen Cloth oiled or varnish'd, and many other Substances soaked in such Liquors as will intimately pervade their little Pores, become by that means more transparent than otherwise; so, on the contrary, the most transparent Substances, may, by evacuating their Pores, or separating their parts, be render'd sufficiently opake; as Salts or wet Paper, or the *Oculus Mundi* Stone by being dried, Horn

> by being scraped, Glass by being reduced to Powder, or otherwise flawed.[18]

Thus, Newton argues that an opaque object can be made transparent by filling in the gaps between those constituent particles, for example by soaking paper in oil, which reduces the amount of refraction that is experienced as light travels from least part to least part.

Returning to Fitz James O'Brien's description of invisibility, we can now see that he likely got his ideas from Newton's pioneering research, or at least from someone else's description of it. O'Brien suggested that "a certain chemical coarseness is all that prevents its being so entirely transparent as to be totally invisible," and this is in agreement with Newton's hypothesis that opacity is entirely due to imperfections in the material at the smallest level.

In a sense, then, we can trace scientific explanations of invisibility right back to Newton. Though he never hypothesized about invisible monsters, his attempts to understand the nature of transparency would fire the imaginations of science fiction authors and, through them, eventually scientists.

Newton was, however, largely incorrect in his explanation of opacity. As we will see, the interaction of light and matter is generally much more complicated and requires an understanding of the nature of atoms themselves, something that Newton had no ability to study. His explanation does work for paper, however, which consists of a bunch of transparent fibers woven together with air gaps between them. Light gets blocked from passing through the paper by this forest of fibers, much like the Vantablack material traps light in a forest of carbon nanotubes. Soaking a piece of paper in oil fills in the gaps between the fibers in a way that reduces the refractions and reflections, resulting in transparency.

Fitz James O'Brien's explanation of invisibility gives us a snapshot of optical science in his era: relatively little was known about the behavior of light, and almost nothing about the properties of atoms. This situation would change dramatically over a few short decades and lead to new visions of invisibility previously unimagined.

4
Invisible Rays, Invisible Monsters

The bushes were now quiet, and the sounds had ceased, but Morgan was as attentive to the place as before.

"What is it? What the devil is it?" I asked.

"That Damned Thing!" he replied, without turning his head. His voice was husky and unnatural. He trembled visibly.

I was about to speak further, when I observed the wild oats near the place of the disturbance moving in the most inexplicable way. I can hardly describe it. It seemed as if stirred by a streak of wind, which not only bent it, but pressed it down—crushed it so that it did not rise, and this movement was slowly prolonging itself directly toward us.

Ambrose Bierce, "The Damned Thing" (1893)

Some fifty years before Fitz James O'Brien pondered the possibilities of the imperceptible, scientists had already discovered an unseen world, though of a form unexpected and unimagined. At the beginning of the nineteenth century, researchers discovered two types of invisible light that cannot be seen with the naked eye but nevertheless influence the material world. The idea that there is an entire universe of physics hidden from our direct perception would not only inspire science fiction authors to present new ideas of invisibility but later influence the science of invisibility in truly surprising ways.

These discoveries would be a key piece of the puzzle in attempting to answer the question, "What is light?"

The first discovery on this path was a serendipitous finding made in 1800 by the German-born British astronomer—and musician—William Herschel (1738–1822). Born in the Electorate of Hanover in Germany, Herschel seemed poised to follow in the footsteps of his father, Isaac, as an oboist in the Hanoverian military. In 1755, William (with his brother Jakob) was stationed in England, giving him his first exposure to British life; at that time, King George II had united the crowns of Great Britain and Hanover. But war with France brought the brothers back to Hanover to defend their homeland, where the Hanoverian forces were defeated in the Battle of Hastenbeck in July 1757.

The possibility of losing his sons was too much for Isaac Herschel, who sent the young men to seek refuge in England later that year. William lived for many years on his musical talents, both as a composer and a performer. He learned to play violin, harpsichord, and organ and worked at times as an orchestral soloist as well as a church organist.

Even in his musical efforts, Herschel demonstrated a knack for problem solving that would serve him well in his future scientific efforts, as the following anecdote from an organist named Miller shows.

> A new organ for the parish church of Halifax was built about this time, and Herschel was one of the seven candidates for the organist's place. They drew lots how they were to perform in succession. Herschel drew the third, the second fell to Dr. Wainwright of Manchester, whose finger was so rapid that old Snetzler, the organ-builder, ran about the church exclaiming: "Te tevel! te tevel! he run over te keys like one cat; he will not give my piphes room for to shpeak." "During Mr. Wainwright's performance," says Miller, "I

was standing in the middle aisle with Herschel. What chance have you," said I, "to follow this man?" He replied, "I don't know; I am sure fingers will not do." On which he ascended the organ loft, and produced from the organ so uncommon a fulness, such a volume of slow, solemn harmony, that I could by no means account for the effect. . . .

"Ay, ay," cried old Snetzler, "tish is very goot, very goot indeet; I vil luf tish man, for he gives my piphes room for to shpeak." Having afterwards asked Mr. Herschel by what means, in the beginning of his performance, he produced so uncommon an effect, he replied, "I told you fingers would not do!" and producing two pieces of lead from his waistcoat pocket, "one of these," said he, "I placed on the lowest key of the organ, and the other upon the octave above; thus by accommodating the harmony, I produced the effect of four hands, instead of two."[1]

Though professional musician to scientist may seem to be an unusual life transition, Herschel's musical training not only prepared him for the change but propelled him along it. He was educated from an early age in a range of mathematical topics because his father wanted him to understand the theory of music as well as its practice. Later in life, around the year 1770, Herschel turned again to science to improve his musical skills. He read Robert Smith's *Harmonics; or, The Philosophy of Musical Sounds,* and this led him to Smith's *Compleat System of Opticks,* a comprehensive book that not only covers the theory of light and imaging but provides practical techniques for telescope design as well as observations of the Sun, Moon, and other heavenly bodies known at the time. It appears that reading *Opticks* excited a curiosity in Herschel to see for himself all the wondrous things that Smith had described, and he set himself on the path of becoming an astronomer.[2]

Around the same time, William made another choice that would turn out to be key to his future scientific career. His younger sister Caroline Herschel (1750–1848) had remained in Hanover, where she seemed destined for an unremarkable life. Her mother had restricted Caroline's education to the most basic and practical skills, apparently fearing that Caroline might become too educated and leave home like her older brothers. In 1772, however, William invited Caroline to join him in England to work as a singer in the choir where he performed. He came to get her in that same year, just as his interest in the stars was evolving into an obsession. Caroline had success as a singer but also ended up attending to her brother as he worked the telescope. She took notes while he observed and made sure he ate his meals on time. Her role evolved in a few short years from servant to laboratory technician, aiding William in the construction of increasingly larger and more sophisticated telescopes, to astronomer in her own right. She ended up discovering eight comets during her career, and her work led to her being made an honorary member of the Royal Astronomical Society and the first woman in England to be honored with an official government position.

William started looking at the heavens in 1773, though his serious astronomy work began in 1779 with a systematic search for stars very close together visually in the sky. It was believed at the time that studying the apparent change in position of such double stars could be used to deduce their motion through space as well as their distance from Earth. In March 1781, while performing his survey of double stars, Herschel observed a previously unknown disklike object in the sky. He had made the first observation of the planet Uranus, and this discovery not only cemented his fame as a first-class astronomer but also earned him the prestigious position of Fellow of the Royal Society.

In February 1800, William had initiated a series of direct telescope observations of the Sun. Such observations require the sunlight to be heavily attenuated in order to be viewed safely, and Herschel experimented with a variety of glasses of different colors to filter the incoming light. As he later wrote, "What appeared remarkable was, that when I used some of them I felt a sensation of heat, though I had but little light; while others gave me much light, with scarce any sensation of heat."[3] Herschel noted, in particular, that a red-colored filter blocked a lot of light but seemed to allow a great amount of heat to pass through it. Everyone was aware that sunlight provides heat, but it now occurred to Herschel that different colors of light might provide different amounts of heat and also that certain colors might be more effective at providing illumination than others.

He embarked on a series of experiments to determine the heating and illumination powers of all the colors of light. He used a prism set up by a window to create a rainbow pattern on a table. To test illumination, he passed a microscope through the rainbow and judged by eye the visibility of objects when lit with different colors. To test heating, he placed a set of thermometers with their bulbs situated in successive colors (fig. 6).

Herschel found, as one might expect, that objects appeared most visible when illuminated with yellow light, in the center of the spectrum, and less visible on the extreme ends of the spectrum with red and violet light. But in heating, he found that violet light produced almost no perceptible heating at all, and heating increased continually as one approached the red end of the spectrum, where it was greatest.

At this point Herschel made a remarkable deduction that would revolutionize our understanding of light. Recognizing that "radiant heat" appears to be subject to the same laws of refraction and dispersion as visible light, he noted, "May not this lead us to surmise, that radiant heat consists of particles of light of a certain range of momenta, and which range—may extend a little farther, on each side of

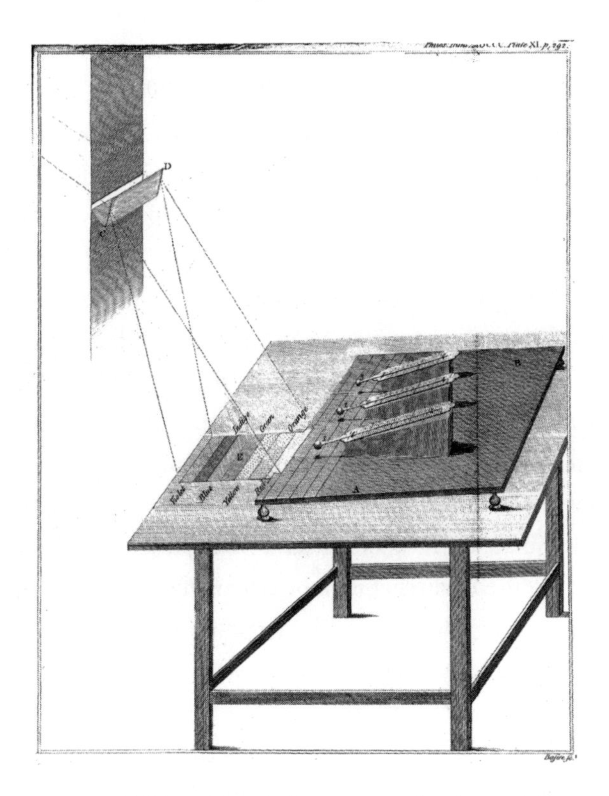

Figure 6. William Herschel's apparatus for determining the heating power of both visible and invisible rays. Illustration from his paper "Experiments on the Refrangibility of the Invisible Rays of the Sun" (1800).

refrangibility, than that of light?"[4] There are two huge intuitive leaps in this statement. First, it had long been assumed that the heat of sunlight was a distinct phenomenon from the illumination itself, and here Herschel suggests that radiant heat is a form of light itself or perhaps a by-product of it. Second, Herschel notes that, because the heat seems to increase all the way to the end of the visible spectrum, it may increase further just outside of this spectrum, where no light is detected.

He tested this latter hypothesis in a follow-up paper in 1800, "Experiments on the Refrangibility of the Invisible Rays of the Sun." Now, instead of placing the thermometers in a visible color of sunlight, he placed them just outside of the range of reds—and found that the temperature of the thermometers went up even more than before.

Herschel had discovered what is now known as infrared radiation, which is not perceptible to the human eye but is exceedingly good at heating matter. He referred to these as "calorific rays" and initially assumed, correctly, that the source of illumination and heat in the full spectrum of rays, both visible and invisible, are two aspects of the same phenomenon: "To conclude—if we call *light,* those rays which illuminate objects, and *radiant heat,* those which heat bodies, it may be inquired, whether light be essentially different from radiant heat? In answer to which I would suggest, that we are not allowed, by the rules of philosophizing, to admit two different causes to explain certain effects, if they may be accounted for by one."[5] Herschel was not a physicist by training, and he seems to have backed away from his conclusion in later work, deciding that heat and light are different. His original assumption, however, would later be fully vindicated. The reason that some rays of the spectrum produce much illumination and little heat, and vice versa, would eventually be explained by a better understanding of how light interacts with matter, as we will see.

In his experiments, Herschel strongly suspected that there must be invisible rays beyond the violet end of the spectrum as well, and he searched for rays with strong heating properties there, without success. But his work inspired the German scientist and philosopher Johann Wilhelm Ritter (1776–1810) to continue the search, and Ritter would find evidence of additional invisible light that would eventually come to be known as ultraviolet light.

Ritter, a self-taught scientist, was a proponent of the German Naturphilosophie movement that began in the late 1700s. With an emphasis on intuition rather than experiment to explain the great problems of science, Naturphilosophie was very much a reaction to the empirical science of the previous century. The "intuition" of Naturphilosophie viewed all of Nature as a unified whole, from the elementary laws all the way through to living organisms and the workings of the human mind. The view of speculation as superior to experiment would—justifiably—fall out of favor in the mid-nineteenth century, but ironically it would stimulate a surprising number of successful experimental efforts along the way. One can see the influence of the movement to some degree in the modern attempts to create unified theories of physics, in which all observed phenomena can be explained by a single fundamental force.

Naturphilosophie also emphasized the concept of polarity in nature: just as there are positive and negative electric charges, and positive and negative magnetic poles, one could—with a bit of imagination—see polarity in everything. Water molecules, which are necessary for life, come from the combination of two very different elements: oxygen and hydrogen. The air we breathe, also necessary for life, is the combination of oxygen and nitrogen, gasses with very different properties.

With this in mind, one can see how Ritter was inspired to search for invisible radiation beyond the violet end of the spectrum. If calorific rays provide heat beyond the red end, might there not be some complementary rays beyond the violet? Ritter, in particular, imagined that the violet end of the spectrum might have cooling rays that counterbalanced the heating rays at the red end. His reasoning, and the guesswork of Naturphilosophie, would turn out to be incorrect, but it nevertheless led him to a groundbreaking discovery.

Failing to find the cooling rays he predicted, Ritter looked for other light-based phenomena that might be enhanced beyond the violet. It was well known in his time that certain chemical processes are influenced or driven by light, and so Ritter studied the effects of different colors of light on chemistry. In particular, silver chloride was known to change color from white to black when exposed to sunlight. When Ritter exposed silver chloride to the space beyond the violet end of the spectrum, where no visible light was present, he found that the chemical still changed color, and faster than it did when exposed to any visible light. Just as heating was greatest in the infrared portion of the spectrum, chemical processes happened most rapidly in the ultraviolet end of the spectrum. Ritter had demonstrated a different set of invisible rays, which he called "actinic rays."

Today, the concept of infrared and ultraviolet light is everyday knowledge. Infrared radiation is given off by all warm bodies and can be measured with thermal imaging cameras. Our current global warming crisis is driven by the trapping of infrared radiation by greenhouse gases in the atmosphere. Visible light from the Sun passes through the atmosphere freely, where it is absorbed by the Earth and converted into heat; the heated Earth then radiates infrared radiation, which cannot escape back into space due to the presence of carbon dioxide and other gases acting as a barrier. Ultraviolet radiation causes sunburns, and we protect against it by putting on sunblock whenever we go to the beach.

A better understanding of how invisible rays fit into the bigger picture of physics would take many years to unravel. In the short term, however, their discovery was the revelation of a hidden world of phenomena previously unimagined. If there exist forms of light that cannot be seen, why not forms of invisible matter, too?

Science fiction and horror authors were, again, the first to think about how the new physics might be used to achieve or explain invisibility. The first author to use invisible rays in such a manner was author, journalist, soldier, and satirist Ambrose Bierce (1842–1914). Born in Ohio and raised in Indiana, Bierce grew up in a poor household but cultivated a love of reading and writing thanks to his literature-loving parents. At age fifteen, he left home and became a printer's apprentice at a small abolitionist newspaper. When the Civil War broke out, Bierce joined the Ninth Indiana Cavalry of the Union Army and participated in a number of battles in the war, including the bloody Battle of Shiloh of April 1862, where there were more than ten thousand casualties on each side of the conflict. His experiences would inspire and inform a number of later war stories and a memoir.

After suffering a brain injury at the Battle of Kennesaw Mountain in June 1864, Bierce stepped away from the army. He briefly returned to service in 1866, when he joined an expedition to inspect military bases across the West. He ended up in San Francisco, where he began a career in journalism that would occupy most of his life.

The most famous period in Bierce's journalism career, and one that gives an indication of his personality, came in 1896. The Union Pacific and Central Pacific railroad companies had received massive low-interest loans to build the railroad, totaling $130 million (some $4 billion in today's dollars), and Central Pacific executive Collis Potter Huntington traveled to Washington, D.C., in a bid to convince Congress to forgive the remaining $75 million debt, essentially turning the loan into a gift. William Randolph Hearst, owner of the *San Francisco Examiner,* was opposed to this shady deal, and he sent Bierce to Washington to draw attention to it and generally make Huntington's life a living hell. The *Examiner* announced Bierce's assignment with supreme confidence. "Mr. Bierce is the most formidable opponent the railroad monopoly ever had in California, and we have faith that whether we win or lose in Congress he can give Mr. Huntington and the Congressmen that enlist in his camp a liberal installment of their desserts."[6] This statement turned out to be accurate. Bierce's reports shined a bright light on a scheme that Huntington was hoping would pass unnoticed, and Bierce wrote up the proceedings in Congress with a devastating wit and savageness that pressured not only Huntington but his congressional accomplices. The attempt to forgive the railroad debt failed, and Bierce was viewed as a hero.

Bierce allegedly got word early on that Huntington was interested in paying him off to go away quietly. Bierce reportedly replied, "Please go back and tell him that my price is about seventy-five million dollars [the debt that was currently owed]. If, when he is ready to pay, I happen to be out of town, he may hand it to my friend, the Treasurer of the United States."[7]

Bierce wrote 249 short stories during his career, with many of his most famous stories appearing in the 1880s and 1890s. This includes "An Occurrence at Owl Creek Bridge," a surreal war story about a captured spy on the verge of execution, and "An Inhabitant of Carcosa," a supernatural story of loss and despair.

Invisibility was featured in Bierce's story "The Damned Thing," first published in 1893 and quoted at the beginning of this chapter. In the story, told in four parts, an inquest is opened into the brutal and mysterious death of hunter Hugh Morgan. Morgan was slain in full view of a witness, though no assailant could be seen. The possible explanation of the mystery comes in the fourth part of the story, taken from Hugh's own journal, in which he describes his encounters with a supernatural being that he refers to only as the Damned Thing.

> It is known to seamen that a school of whales basking or sporting on the surface of the ocean, miles apart, with the convexity of the earth between them, will sometimes dive at the same instant—all gone out of sight in a moment. The signal has been sounded—too

> grave for the ear of the sailor at the masthead and his comrades on the deck—who nevertheless feel its vibrations in the ship as the stones of a cathedral are stirred by the bass of the organ.
>
> As with sounds, so with colors. At each end of the solar spectrum the chemist can detect the presence of what are known as "actinic" rays. They represent colors—integral colors in the composition of light—which we are unable to discern. The human eye is an imperfect instrument; its range is but a few octaves of the real "chromatic scale." I am not mad; there are colors that we can not see.
>
> And, God help me! the Damned Thing is of such a color![8]

Bierce imagines that a being has evolved to be made of a substance whose color is outside the visible spectrum; just as one can imagine objects that are red in color or blue or green, the Damned Thing is evidently of an infrared or ultraviolet color.

In Bierce's time, science did not know enough about the interaction of light and matter to rule out this idea completely, though this would change in only a few short years.

Ambrose Bierce would pull off a disappearing act of a very different sort in his own life. In 1913, then seventy-one years of age, he traveled to Mexico to observe the Mexican Revolution. In Ciudad Juárez he met up with the army of revolutionary general Pancho Villa to join them as an observer. He accompanied the soldiers to the city of Chihuahua, where he sent a final letter to a friend. And after that . . . he was never heard from again. No evidence of his ultimate fate has ever been uncovered.

Bierce's story of an invisible monster was likely influenced by an earlier story by one of his contemporaries, French author Guy de Maupassant (1850–1893). Maupassant was a prolific master of the short story form, and "The Horla," first published in 1886 and expanded into its final form in 1887, is one of his most famous.[9]

The story is conveyed as the diary of an unnamed narrator who, despite a comfortable and happy life in Paris, begins to suspect that he is under the psychic influence of an invisible being that he calls the Horla (fig. 7). When trying to rationalize the existence of his unseen tormentor, the narrator recalls the words of a monk he had conversed with: "Can we see the hundred-thousandth part of what exists? Listen; there is the wind which is the strongest force in nature; it knocks men down, blows down buildings, uproots trees, raises the sea into mountains of water, destroys cliffs, and casts great ships on to the breakers; it kills, it whistles, it sighs, it roars,—have you ever seen it, and can you see it? It exists for all that, however!"[10] Maupassant does not reference the invisible colors of light but does hint at it with mention of things that cannot be seen in nature. And the Horla is not a supernatural being: it eats and drinks the narrator's food when he is asleep. One can see how Bierce may have been inspired by Maupassant's tale and wrote his own take with a more scientific explanation.

"The Horla" is a story with distinctly apocalyptic overtones. The narrator regards the creature as the next stage of evolution, a being superior to humanity that will supplant it. Late in the tale, the narrator learns that a seeming madness has swept Rio de Janeiro, with residents fleeing their homes in fear of invisible invaders. He then recalls waving to a Brazilian ship passing his home at the beginning of the diary and realizes that he had inadvertently welcomed the monster into his home. In the end, he takes drastic action to trap and destroy the creature before it overwhelms his mind completely.

Guy de Maupassant suffered from psychological problems later in life, including paranoia and a phobia of death. It has been suggested that "The Horla" is a reflection of Maupassant's own struggles. He was institutionalized in 1892 and died in 1893. His death reverberated around the world, and the literary world mourned the loss of one of their own. His friend the novelist Émile Zola gave remarks at

Figure 7. The Horla tormenting its victim. Illustration from Guy de Maupassant, *Works* (1911).

his burial that summarized the complexities of his life well: "Outside of his glory as a writer, he will remain as one of the men who have been the most fortunate and the most unfortunate that have lived, the man in whom we feel the hope and the despair of humanity, the adored, spoiled brother disappearing amid tears."[11]

5

Light Comes Out of the Darkness

> Completely invisible, traveling many miles per second, his ship headed to Mars! He must have hurtled through mine fields, but that didn't matter now. The devouring disintegration rays that poured out from the walls of his great machine ate up mines before they could explode, and simultaneously destroyed every light wave that would have revealed his craft to alert eyes out there in the blaze of sun.
>
> *A. E. van Vogt,* Slan *(1946)*

At the turn of the nineteenth century, humanity's understanding of the nature of light underwent a dramatic transformation, one that would eclipse even the discovery of invisible light rays. For nearly one hundred years, the study of optics had been dominated by the views of Isaac Newton, who published his classic work on the subject, *Opticks,* in 1704. Newton's work seemed to put to rest an argument that had raged in his time—is light a stream of tiny particles or a wave, like water and sound? Newton did rigorous experiments, testing the properties of light in every way imaginable in his era, and concluded from his efforts that light consists of a stream of particles.

There were a few strange phenomena that Newton's theory could not yet explain. For instance, in 1665, the Italian Jesuit priest Francesco Grimaldi noticed that a narrow beam of light appeared to spread out after passing through a small slit in an opaque screen. He referred to this phenomenon as diffraction, from the Latin *diffringere,* "to break into pieces." Researchers did not view diffraction as a major problem for Newton's theory, seeing it as a minor puzzle that would eventually be solved using the Newtonian system.

In 1800, however, the British scientist Thomas Young published the first of several papers arguing that light does, in fact, have wave-like properties, and his research would be the start of a new era of wave optics that continues to this day.

Born in the village of Milverton in Somerset, England, in 1773, Thomas Young stood out even as a young child. He could read fluently by age two, and by age four he had read the Bible twice through. He demonstrated incredible proficiency in languages, and by his teens he had translated parts of the Bible into thirteen different languages. At age fourteen, he even took on a role as a tutor for a family friend.[1]

Language was not Young's only passion, however. He read widely, including books on natural philosophy, and became "particularly delighted" with the study and practice of optics.[2] In his teens, with the help of a staff member at his school, Young learned how to design and build telescopes.

Though he had diverse scholarly interests, Young initially focused on a career as a medical doctor. He was inspired by his great-uncle Richard Brocklesby, a physician of high renown in London. Brocklesby had treated Young when he fell seriously ill in his teens, saving his life. But Young was motivated by more than gratitude: by pursuing the medical path, he was guaranteed an inheritance from his uncle, and financial security with it. So in 1793, Young became a student at the venerable Saint Bartholomew's Hospital in London, established in the year 1123.[3]

Even in medicine, Young found himself drawn into questions of optics. In his studies of anatomy, he had learned of an unsolved question in vision: How does the eye of a living creature adjust, or accommodate, to produce clear images of objects at any distance? Through dissection of an ox eye, Young came to the conclusion that the crystalline lens of the eye has its shape distorted by the actions of muscles, changing the focusing properties of the lens accordingly. Young produced a paper on the subject and presented it to the prestigious Royal Society in London on May 30, 1793. The paper was initially so well received that Young was made a Fellow of the Royal Society the following year, at age twenty-one.

Young's paper, "Observations on Vision," soon met with controversy and condemnation.[4] Research rivals argued that Young's results were incorrect because they themselves had not observed any deformation of the lens of the eye in their investigations. Adding to Young's troubles, a famous surgeon named John Hunter claimed that Young had overheard him discussing the eye and copied his ideas on the subject. Though the plagiarism charge was quickly dismissed, Young ended up renouncing his work on the human eye for a time, deferring to the experts in the field. This retraction would cause him additional troubles in the future, though science has since shown that his explanation of accommodation was correct.

As part of his plan for a well-rounded medical education, Young proceeded to Edinburgh to continue his studies and from there went to Göttingen to earn a degree. He did not stay in Göttingen for long, however: he and his uncle had misunderstood the rules for practicing medicine in London, which required two years of residence in a London medical school to become a Fellow of the Royal College of Physicians. On learning this, Young hastened to finish his degree in Göttingen and ended up staying for only nine months.

As part of the degree requirements in Göttingen, Young had to give a lecture related to medicine. He chose as his topic the working of the human voice. This required him to study the properties of sound waves, and during his investigations he was struck by the similarity of phenomena involving sound and those involving light. Though researchers had long concluded that light does not possess wave properties, the similarities between light and sound seemed too strong to be coincidental, and they drove Young to explore the possibility that light is in fact a wave.

After Göttingen, Young entered Emmanuel College, Cambridge, for the last stage of his training, and finished two years later, in the autumn of 1799. After that, he set himself up in private medical practice in London, as planned. Such practices were slow to grow and accumulate patients in that era, however, leaving Young plenty of free time to ponder those scientific questions that had puzzled him for years.

He began his efforts with a number of essays on a range of scientific topics, including the properties of sound waves, that appeared in the *British Magazine* under the pseudonym "The Leptologist." Having been embarrassed in public while sharing his scientific views in the past, he appears to have used the pseudonym as a way to wade back into debate without risking his reputation. He officially reentered the scientific discourse in a letter that was published by the Royal Society in January 1800, "Outlines of Experiments and Inquiries respecting Sound and Light."[5] The letter is primarily an analysis of sound waves and their behavior, but Young also noted the similarities between sound and light, a hint of his work to come. Later that year, Young published "On the Mechanism of the Eye" in the *Philosophical Transactions of the Royal Society*, a renewed defense of his hypothesis on the properties of the lens of the eye.[6]

To further his renewed scientific activity, Young accepted a job in 1801 as a professor of natural philosophy at the Royal Institution, an organization that had been founded only two years earlier to foster scientific education and research. With this new role, he focused his energies on studying the properties of sound and their remarkable similarities to light. To understand what Young saw, we now need to spend a little time discussing what, in fact, a wave is! In technical terms, we may describe a wave as an oscillatory motion of "something" that transports energy from one place to another but does not transport the "something" itself. Waves in water are the simplest example to visualize because they move slowly enough to be seen with the eye and also possess all of the properties of other waves, such as sound waves and light waves.

When a rock is dropped into a pond, ripples spread out from the point of impact, usually manifesting as a succession of up-and-down regions where the water level is higher or lower than normal, respectively. These ripples can travel great distances along the surface of the water before dissipating and can disturb objects on the water's surface, such as leaves (or waterfowl). This ability to move distant objects demonstrates that the waves carry energy. There is no net flow of water away from the place the rock is dropped itself: the water level (the "something" for water waves) locally goes up or down, but the water itself does not move away from the point of impact. This behavior is distinct from the motion of water in a stream, where the water actually flows downstream en masse and eventually empties into a lake or an ocean.

Another example of a wave is a vibration on a stretched piece of string or elastic, such as a string on a guitar (or a classic Slinky toy). When the guitar string is plucked, that disturbance is carried along the length of the string. The waves carry energy along the string, but the string itself stays in one place, firmly attached to the guitar.

The simplest form of wave is one that consists of a simple repeating up-and-down motion, mathematically taking on the form of a sine wave from trigonometry (fig. 8). In optics, for reasons that will become clear, these are called monochromatic (single color) waves.

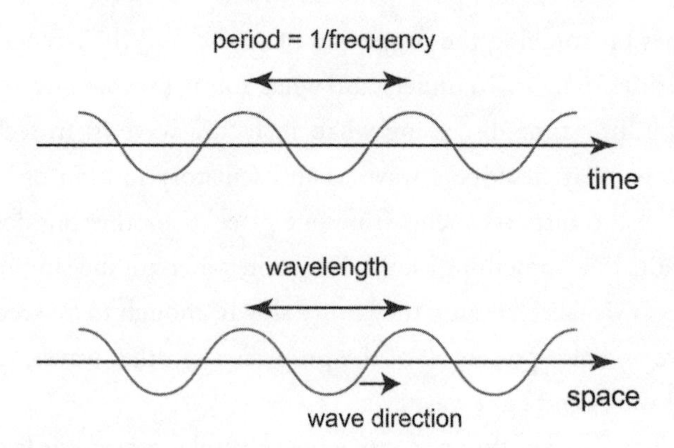

Figure 8. Monochromatic waves on a string.

If we look at the string at a single point along its length and watch what it does in time, we see that the wave will alternately move upward and downward. This is analogous to sitting in a docked boat and feeling the boat rise and fall beneath you as waves pass by. The amount of time between peaks is called the period of the wave, and the inverse of this is called the frequency, which tells you how many peaks you encounter every second.

If we take a snapshot of the entire string at a single point in time, we see a similar picture: along its length, it has alternating upward and downward wiggles. The spacing between the peaks in space is called the wavelength: it represents the physical length of a single up-and-down cycle of the wave.

For sound waves, the "something" that is waving is the density of the air molecules. These alternating regions of high and low density travel through the air and cause vibrations on our eardrums that we perceive as sound. Again, this is a motion distinct from the transport of air molecules through space, which we perceive as a wind or a breeze. In music, middle C has a frequency of 261 cycles per second,

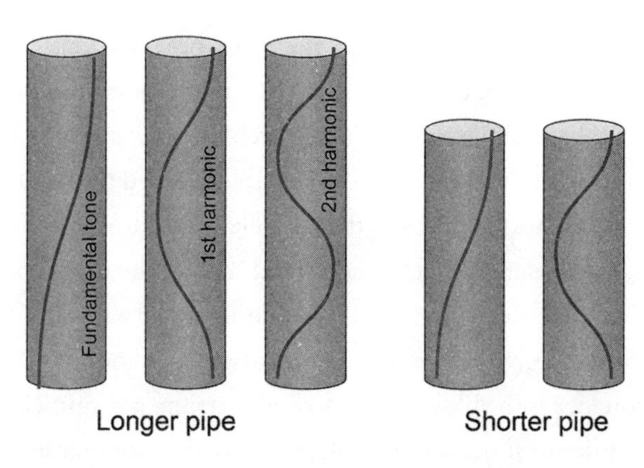

Figure 9. Resonant waves in an organ pipe, for longer and shorter pipes.

which corresponds to a wavelength of 132 centimeters. Higher notes have higher frequencies and correspondingly shorter wavelengths.

One particular property of sound waves drew Young's attention, a phenomenon known today as resonance. When a sound is created in an enclosed space, waves with wavelengths that fit perfectly into that space will build on themselves, becoming louder. If you've ever thought your singing sounds better in the shower, you've experienced the resonance of sound waves: the walls of the shower form an enclosed space that causes particular tones to be enhanced. A resonant wave doesn't actually move but oscillates in place in its confined space and is referred to as a standing wave.

A simple example of resonance is found in the operation of pipe organs. An organ pipe with openings at both ends naturally fits waves inside it that have a maximum or minimum at each end. This means that the lowest tone that a pipe will produce will have a half wavelength fitting inside the pipe (fig. 9). A wave of this length will build up rapidly over time, producing a loud, clear tone. Waves of shorter wavelengths will also resonate in the pipe, provided that they, too, are of such a wavelength that there is a maximum or minimum at each end. This means that a number of different wavelengths can resonate in the pipe; the longest wavelength possible is called the fundamental tone, and the shorter wavelengths are called the harmonics. The fundamental tone is usually the loudest and defines the distinctive pitch of the note. The combination of the fundamental tone and the harmonics is what gives an instrument its distinctive sound.

Resonance is the basis for most wind instruments: musicians playing a recorder or flute, for example, change the note they are playing by opening or closing holes in the instrument, effectively making it a shorter or longer hollow tube. Brass instruments like the trumpet and tuba use valves to redirect the sound through longer or shorter paths, changing the resonance frequency accordingly.

In the resonant behavior of organ pipes, Young saw an explanation for an optical phenomenon discussed in detail by Isaac Newton in his *Opticks* and later named after him: Newton's rings. Newton placed a glass lens with a very shallow curvature on top of a flat piece of glass and observed colorful rings emanating from the center outward. These rings are very small but can be magnified or observed in a microscope (fig. 10).

Newton interpreted the colored rings as the result of a complicated process of reflection and refraction of light as it bounces back and forth inside the gap between the lens and the glass plate. Young, however, saw something else: if light is a wave, then the gap, which increases in thickness as one moves away from the center, could serve as a sequence of organ pipes for light. The colors seen by Newton would therefore represent light waves resonating between the pieces of glass in regions of different gap size. Each color of light would represent a wave with a different wavelength and frequency. Looking at a classical pipe organ and visualizing little pipes in the gaps in New-

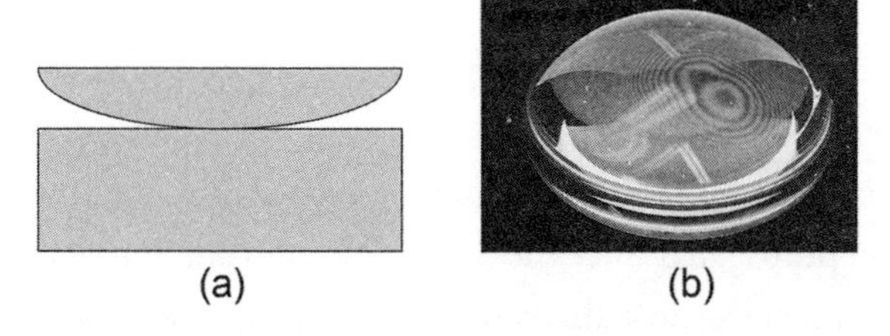

Figure 10. (a) Side view of Newton's experimental setup (thickness of lens exaggerated); (b) rings observed from above, magnified by the upper curved surface. Photo: Ulfbastel / Wikimedia Commons / Public Domain.

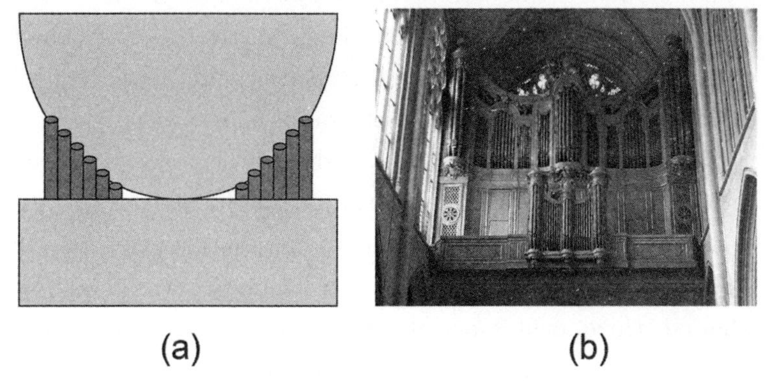

Figure 11. (a) Newton's rings experiment with imaginary pipes; (b) the pipe organ in Saint-Germain l'Auxerrois, Paris. Photo: Gérard Janot / Wikimedia Commons / Public Domain.

ton's rings experiment, one can imagine how Young came to his inspired idea (fig. 11).

Young first hinted at these possibilities in his letter "Outlines of Experiments and Inquiries respecting Sound and Light." In the short section on light, "The Analogy between Light and Sound," he presented the similarity between Newton's rings and pipe organs and reintroduced the idea, first given by the mathematician Leonhard Euler, that the colors of light are the visible manifestations of the light's frequency. Red light has the longest wavelength, and violet light has the shortest.

Young also addressed a common misconception about waves that had been propagated since the time of Newton and was part of Newton's argument against the wave nature of light: that waves, when generated, spread equally in all directions. Newton noted that this was true for known waves, such as those created by dropping a rock into a pond. Light, however, was observed to be highly directional, as when sunlight peeks through dark clouds, creating what are technically known as crepuscular rays but are often called god rays.

Young argued that Newton was simply mistaken: though sound waves do spread out a lot, they can also be quite directional. He provided several examples, with a subtle bit of humor added in: "It is well known, that if a person calls to another with a speaking trumpet, he points it towards the place where his hearer stands: and I am assured by a very respectable Member of the Royal Society, that the report of a cannon appears many times louder to a person towards whom it is fired, than to one placed in a contrary direction."[7] There was reason, then, to think that light waves could also travel without significantly spreading. Light waves would then not even need "pipes" like an organ in order to produce the colors of Newton's rings.

On November 12, 1801, Young gave a prestigious Bakerian Lecture to the Royal Society, "On the Theory of Light and Colours."[8] In it, he provided a comprehensive theory of the wave properties of light. Among his achievements, he estimated the wavelength and frequency of the different colors of light, using Newton's original measurements of the gap thicknesses in Newton's rings. For red light, for example, he calculated a wavelength of 0.675 billionths of a meter and a frequency of 463 million-million oscillations per second. For blue light, he calculated a wavelength of 0.5 billionths of a meter and a frequency of 629 million-million oscillations per second. These numbers are quite reasonable by modern standards considering that the terms "red" and "blue" refer to a range of frequencies and wavelengths. The high frequency and small wavelength of light waves also explained in part why people had not noticed the wave properties of light before: the oscillations are too fast and too small for the human eye to detect in ordinary circumstances.

It was not just courtesy that motivated Young to use Newton's own measurements and experimental observations: he well knew that criticizing the legendary man's work could provoke a strong backlash against himself, the young upstart. Young took great pains in his paper to point out how Newton's experiments and hypotheses could be used to support a wave theory of light. Young was, in essence, attempting to give Newton partial credit for the discovery.

In the myriad hypotheses Young introduced in his paper, it would have been easy to overlook one of these, in which he considered what happens when waves cross paths with each other. He introduced what would eventually be called the law of interference, one of the most important properties of waves, as follows: "When two Undulations, from different Origins, coincide either perfectly or very nearly in Direction, their joint effect is a Combination of the Motions belonging to each."[9] In terms of water waves, we may imagine two waves created separately, intersecting at a point. If both waves are moving upward at the same time, their actions will combine to make an even bigger wave; if one wave is moving up and the other wave is moving down, their actions will at least partially cancel each other out, leading to a smaller wave than either contribution. The former case is now called constructive interference, while the latter case is called destructive interference.

This is illustrated here for two square waves on a string, propagating toward each other (fig. 12). If both wave pulses consist of an

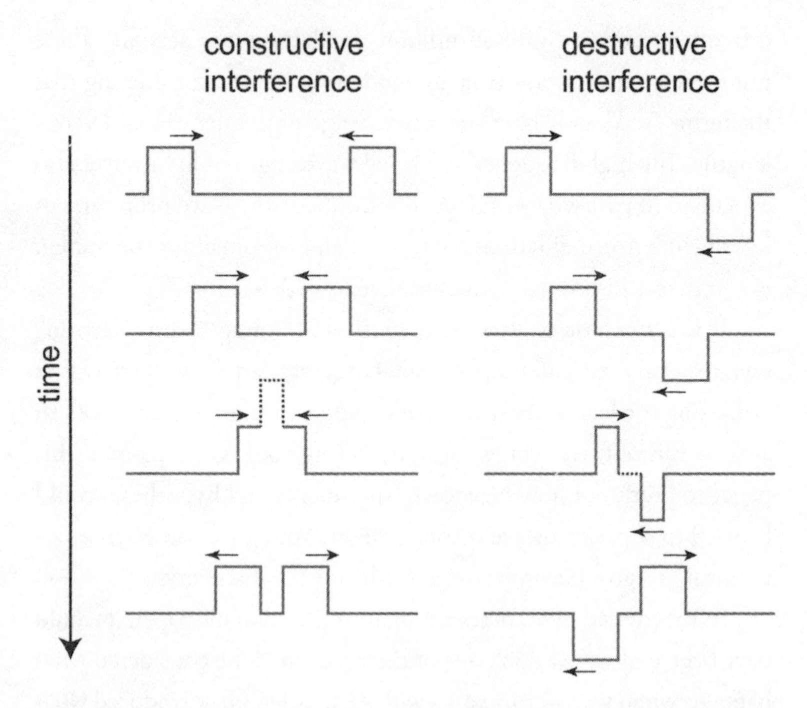

Figure 12. The interference of two waves traveling in opposite directions, when they constructively or destructively interfere. The dashed regions indicate where the waves are overlapping and interfering.

"upward" wiggle, they will grow when they intersect; if one is "up" and one is "down," they will cancel. It is to be noted that the waves don't destroy each other—they carry on their separate ways unchanged once they pass each other. In a word, they "interfere" with each other while they are passing.

Young used his law of interference to explain a number of optical systems where colors unexpectedly appear; these included Newton's rings as well as light scattering from parallel scratches on a polished surface, which today would be referred to as a diffraction grating. If you have ever seen the bright rainbow colors that reflect from the shiny side of a compact disk or Blu-ray disk, you've seen a diffraction grating: the tiny bumps and pits that are used to store the data on the disk act as a grating.

In a follow-up paper to the Royal Society, presented in July 1802, Young introduced a new observation: bands of colors appearing when light goes around a fine fiber or hair.[10] Young cut a small hole in a piece of cardboard and fixed the fiber across the center of the hole. Looking at a distant light source through the hole, he saw colored bands of light on either side of the fiber, parallel to it. He interpreted these colors as arising from interference between light waves that pass on opposite sides of the fiber. Because different colors have different wavelengths, the points in space where their "ups" and "downs" meet are different, resulting in different colors in different locations.

Young continued his work, and in November 1803 he presented his strongest case for the interference of light in a Bakerian Lecture titled "Experiments and Calculations relative to Physical Optics."[11] It would be the first rough demonstration of what later became known as Young's double slit experiment or Young's two-pinhole experiment. In this implementation, Young poked a small hole in a window shutter to allow a thin beam of sunlight to enter his room. In the path of that beam, he placed a thin piece of card, one thirtieth of an inch thick, dividing the beam into two parts, each of which spread into the path of the other. The combined light wave was then projected onto a screen some distance beyond, allowing multiple colored fringes to be seen in the shadow cast by the card. This, Young felt, was conclusive proof of the wave nature of light.

Young refined his experiment over time. The thick card was replaced with "two very small holes or slits" in his later book *A Course of Lectures on Natural Philosophy and the Mechanical Arts* (1807).[12] In this arrangement, the light that forms a pattern on the observation screen comes only from the small holes, and the interference pattern

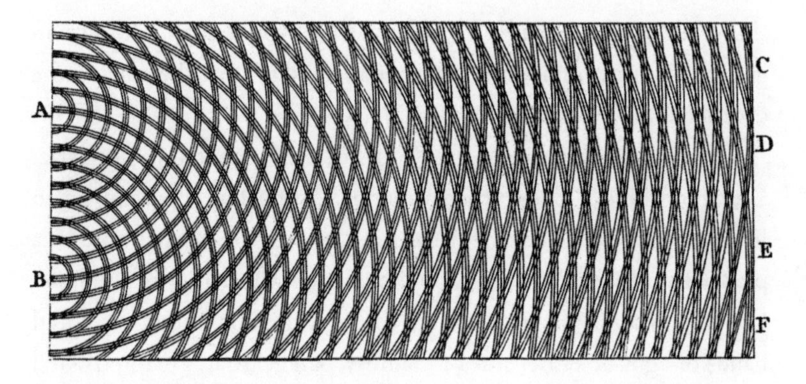

Figure 13. Young's interference experiment. Illustration from Young, *A Course of Lectures on Natural Philosophy and the Mechanical Arts* (1807).

is consequently much easier to see. In this book, Young provided a beautiful picture of how the waves from two holes produce interference (fig. 13).

In this picture, A and B represent the two pinholes. The light waves spread out as circular ripples from the holes and eventually overlap. If we consider the white regions of each ripple as the "up" part of the waves and the dark lines as the "down" part of the waves, one can see that dark regions are formed in the image where the dark lines of one source cross the white regions of the other source. These are regions of complete destructive interference, and they form lines that end at points C, D, E, and F on the observation screen. If a light source consisting of a single color is used to illuminate the holes, the pattern on the measurement screen consists of a sequence of light and dark lines—the regions of constructive interference and destructive interference.

Young made one other significant observation of particular interest. In 1800, the German astronomer William Herschel had discovered infrared radiation, invisible heat radiation lying just outside the red side of the visible spectrum. The following year, the chemist Johann

Wilhelm Ritter discovered ultraviolet radiation, lying outside the violet side of the visible spectrum and detectable through chemical changes. Young reproduced his Newton's rings experiments with ultraviolet radiation and, through the use of paper dipped in a solution of nitrate of silver, acting like photograph paper, showed that ultraviolet light also produces interference. He therefore demonstrated that the wave properties of light extend out beyond one side of the visible spectrum and speculated that the same would be true for infrared light.

It appears that Young's research was at least politely received. He had, as we have seen, been asked to give three Bakerian Lectures within a four-year period, two on his theory of light and one on the mechanism of the eye. But this nonhostile reception did not apparently result in widespread acceptance of his ideas.

Young worked hard to avoid controversy by giving heavy credit to Newton in his work, but controversy found him nevertheless, stemming from an unexpected source—his own injudicious earlier writings. In one of his essays as the Leptologist, Young somewhat flippantly criticized "a young gentleman in Edinburgh" for rediscovering things that had been well known for years. This "young gentleman" was Henry Peter Brougham, who like Young had taken an interest in optics at an early age and had written several papers on the subject for the Royal Society (he would be made a Fellow in 1803). Unlike Young, however, Brougham was a devoted disciple of Newton, and the combination of perceived attacks by Young on himself and Newton was too much to bear. In 1802, Brougham founded the magazine the *Edinburgh Review,* and in 1803 he anonymously launched a series of vicious attacks on Young's optical work in the magazine.

On Young's "Theory of Light and Colours," for example, Brougham began his attack thusly, "As this paper contains nothing which deserves the name, either of experiment or discovery, and as it is in fact destitute of every species of merit, we should have allowed it to pass among the multitude of those articles which must always find admittance into the collections of a Society which is pledged to publish two or three volumes every year."[13] On Young's paper of 1802 discussing the colors produced around fibers, Brougham appeared almost gleeful as he accused Young of rediscovering known facts, something that Young had earlier accused him of: "We are sorry to find that Dr Young is by no means more successful at making observations and experiments, than in forming systems. The new case of colours which he affects to have discovered, has been observed a thousand times; and he has only the merit of giving an absurd and contradictory explanation of it."[14] The attacks, though in fact largely based on errors and misunderstandings, definitely wounded Young's pride. He responded by publishing his own short pamphlet of rebuttals in 1804, *A Reply to the Animadversions of the Edinburgh Reviewers.* From the very first words, it is clear that the argument between the men had gone far beyond a simple scientific dispute:

> A Man who has a proper regard for the dignity of his own character, although his sensibility may sometimes be awakened by the unjust attacks of interested malevolence, will esteem it in general more advisable to bear, in silence, the temporary effects of a short lived injury, than to suffer his own pursuits to be interrupted, in making an effort to repel the invective, and to punish the aggressor. But it is possible that art and malice may be so insidiously combined, as to give to the grossest misrepresentations the semblance of justice and candor.[15]

Young defended not only his scientific ideas but rebuked attacks on his character. Young had retracted, then resubmitted, his ideas on the behavior of the eye as new information came in, and Brougham used this to characterize Young as a man confused about science. Young in return laid out a detailed history of the entire controversy, explaining why he had taken the actions that he had. In the end, though, Young seemed worn down by the controversy that had dogged his scientific efforts from the beginning, and he announced his intention to return to medicine:

> With this work, my pursuit of general science will terminate: henceforwards I have resolved to confine my studies and my pen to medical subjects only. For the talents which God has not given me, I am not responsible, but those which I possess, I have hitherto cultivated and employed as diligently as my opportunities have allowed me to do; and I shall continue to apply them with assiduity, and in tranquility, to that profession which has constantly been the ultimate object of all my labours.[16]

And for a while, other than the publication of his *Course of Lectures on Natural Philosophy,* Young did return to his medical studies. He had resigned his post at the Royal Institution in 1803, and he finally earned the accreditation he needed to become a physician at Saint George's Hospital in 1811. Optics researchers continued to treat light as a stream of particles and not a wave. But Young's theory of light would soon be vindicated in a spectacular way, and it would secure his place as one of the greatest scientists of the nineteenth century.

Young's most famous contribution to optics, the law of interference, was elegantly described years later by François Arago, who in a biographical memoir of Young in 1835 wrote, "Who would not be surprised to find darkness in the sun's rays,—in points which the rays of the luminary freely reach; and who would imagine that any one could suppose that the darkness could be produced by light being added to light!"[17]

Unbeknown to any researchers at the time, Young's discovery of the law of interference was also a great leap forward for the theory of invisibility. Interference showed that light waves, under the right circumstances, can cancel each other out. Eventually, a novel form of interference would be shown to be a key ingredient in the creation of invisible objects in physics.

6

Light Goes Sideways

Curtis Whateley—of the undecayed branch—was holding the telescope when the Arkham party detoured radically from the swath. He told the crowd that the men were evidently trying to get to a subordinate peak which overlooked the swath at a point considerably ahead of where the shrubbery was now bending. This, indeed, proved to be true; and the party were seen to gain the minor elevation only a short time after the invisible blasphemy had passed it.

Then Wesley Corey, who had taken the glass, cried out that Armitage was adjusting the sprayer which Rice held, and that something must be about to happen. The crowd stirred uneasily, recalling that this sprayer was expected to give the unseen horror a moment of visibility. Two or three men shut their eyes, but Curtis Whateley snatched back the telescope and strained his vision to the utmost. He saw that Rice, from the party's point of vantage above and behind the entity, had an excellent chance of spreading the potent powder with marvellous effect.

H. P. Lovecraft, "The Dunwich Horror" (1929)

The wave theory of light would again seem to go into hibernation after Young's efforts, but behind the scenes, others were following in his footsteps and developing a strong mathematical theory to support his observations. These efforts would result in new and surprising insights into the nature of light, and would begin with a contest.

On March 17, 1817, the French Academy of Sciences announced that diffraction, the spreading of light after it passes through a small hole, would be the topic for the biannual physics prize to be awarded in 1819. These Grand Prix events were intended to spur innovation in physics and drive solutions to unanswered questions—and diffraction remained a puzzle. One of the entrants was a twenty-eight-year-old civil engineer named Augustin-Jean Fresnel, who presented a comprehensive wave theory of light, backed up by experimental work.

Fresnel had, like Young, long been interested in questions of optics. He also found himself with leisure time to explore these questions, but from a very different circumstance. Napoleon Bonaparte, who had been exiled to the island of Elba in 1814, escaped in February 1815 with an eye toward reclaiming his throne. Fresnel was one of those who joined the royalist resistance against Napoleon, though he fell ill and was unable to participate directly. When Napoleon became emperor again, Fresnel found himself ostracized, and he was assigned house arrest that he spent at his mother's house. With this forced free time, he was spurred to start his series of optical investigations, which continued even after Napoleon was again dethroned in mid-1815 and Fresnel's job as a civil engineer was restored. Over the next few years, Fresnel requested a leave of absence from his job several times in order to advance his optical studies further.

Fresnel had long found himself fascinated by the possibility that light is a wave. In the course of his investigations, he struck up a correspondence with the French physicist François Arago, who directed him to the work of Thomas Young. Arago pointed out that Fresnel had inadvertently retraced many of the steps that Young had already taken, but he was generally encouraging of the young engineer. When the Academy of Sciences announced diffraction as the prize problem for 1819, Arago convinced Fresnel to submit his mathematical theory of wave optics to the competition.

The prize committee contained some of the greatest minds in physics of the era: François Arago, Pierre-Simon Laplace, Siméon Denis Poisson, and Joseph Louis Gay-Lussac. When Fresnel submitted his prize entry on July 29, 1818, the committee gave it a thorough investigation. Poisson, one of the supporters of the particle theory of light, noted a curious implication of Fresnel's new wave theory that arises when one calculates how light diffracts around an opaque circular disk. According to the law of interference, Poisson showed, one would expect that there should be a bright line aligned with the axis of the disk on the shadowed side, as the waves diffracting from all edges of the disk should constructively interfere on this axis line. For proponents of the particle theory of light, this was an absurd conclusion—how could light appear miraculously right in the middle of a shadow? But Arago did the experiment and confirmed that the bright line is there, just as Fresnel's theory predicted; this spot is now often referred to as the Arago spot.

Fresnel was awarded the Grand Prix for his work, as was announced on March 15, 1819, at a meeting of the academy. This award alone did not convince people that light is a wave, but it was a turning point for the theory's acceptance. The Arago spot showed that the wave theory could not only explain existing experimental observations but predict new phenomena. This drove other researchers to explore the possibilities, and new evidence to support the wave theory piled up, eventually silencing even the strongest proponents of Newton's corpuscular theory. The era of the wave theory of light had begun.

Fresnel would go on to have a stunningly successful career in optical science, as attested by the seven optical concepts that bear his name. His career was cut short when his health began to decline in late 1822 and he was diagnosed with tuberculosis. He scaled back his efforts in order to reduce stress on his body and prolong his life, focusing his work on the development of a lightweight lens that could be used to concentrate the beams of light from a lighthouse. The sort of lens he developed remains in use today and is known as a Fresnel lens. Despite his reduced workload, he died in July 1827, at only thirty-nine years of age.

Though Thomas Young had stepped away from public discussions of optics during the time of Fresnel's diffraction studies, he was still active in private and, in fact, played a key role in the final acceptance of light as a wave. Fresnel had first attempted to explain the mystery of diffraction in 1815, and as we have noted, his correspondence with Arago directed him to write to Young directly. Fresnel contacted Young in part to apologize for inadvertently duplicating his work. Arago, further intrigued by the possibilities of the wave theory of light, visited Young in 1816 with the physicist Joseph Louis Gay-Lussac. The French scientists had brought what they believed to be new results that strengthened Young's theory, but Young noted that he had already explained the phenomenon in question years before. This led to one of my favorite scenes in the history of optics:

> This assertion appeared to us unfounded, and a long and very minute discussion followed. Mrs. Young was present at it, without offering to take any part in it—as the fear of being designated by the ridicule implied in the sobriquet of *bas bleus* makes English ladies reserved in the presence of strangers; our neglect of propriety never struck us until the moment when Mrs. Young quitted the room somewhat precipitately. We were beginning to make our apologies to her husband, when we saw her return with an enormous quarto under her arm. It was the first volume of the Treatise on Natural Philosophy. She placed it on the table, opened the book, without saying a word, at page 387, and showed with her finger a figure where the curvilinear course of the diffracted bands, which were the subject of the discussion, is found to be established theoretically.[1]

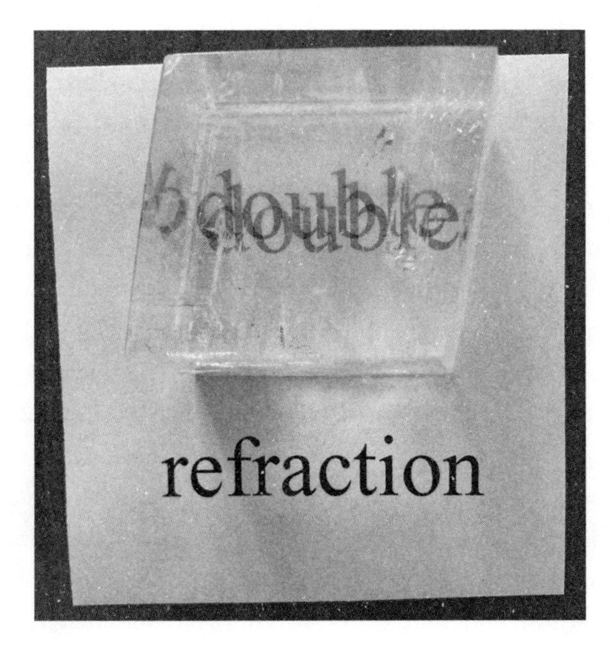

Figure 14. Double refraction through optical calcite.

The visit by Arago and Gay-Lussac brought to Young the latest results related to another important unsolved mystery of light, known as polarization. For centuries, people had noticed that light passing through a clear crystal known as Iceland spar (technically called optical calcite) will produce *two* images of whatever is behind it (fig. 14).

Neither the proponents of light-as-a-wave nor the proponents of light-as-a-particle could provide a satisfactory explanation of this double image, which apparently involves half of the light that passes through the crystal being refracted at one angle and the other half at another angle. The phenomenon is known as double refraction or by the more technical term "birefringence."

An important clue to the nature of double refraction was found in 1808 by the French physicist Étienne-Louis Malus when he happened to look through a piece of Iceland spar at the setting sun, reflected off of the windows of the Luxembourg Palace in Paris. He found, to his surprise, that one of the double images was much brighter than the other and that by rotating the crystal in his hand he could change which image was brightest. He soon found that light reflected off of glass from a certain special angle will produce only a single image in Iceland spar; he referred to light that produces only a single image as "polarized." Light reflected off of a piece of glass lying horizontally would be said to be horizontally polarized, and light reflected off of a piece of glass standing vertically would be said to be vertically polarized. Working together, Fresnel and Arago showed in a series of experiments that mixing vertically polarized light and horizontally polarized light in Young's experiment, with different polarizations illuminating each pinhole, would not produce an interference pattern; somehow, light with different polarizations did not interfere.

Young pondered a possible explanation for these effects, and he began again with an analogy to sound waves. Researchers had found that acoustic waves traveling through a block of wood, Scotch fir in particular, will travel faster when they are propagating along the fibers of the wood than when they are traveling across the fibers. Young, like many researchers, reasoned that something similar must happen in Iceland spar—the crystal must have some interior structure that makes light travel faster along one direction than another, resulting in two different indices of refraction. However, this could not explain why Iceland spar produces two images when light passes through it.

In late 1816, Young finally hit on the solution, and he transmitted it in a letter to Arago on January 12, 1817, marking another major milestone in the understanding of light. Young argued that light is a transverse wave, where the vibrations are perpendicular to the direction that the wave is traveling, in contrast to a sound wave, where the vibrations are along the direction the wave is traveling.

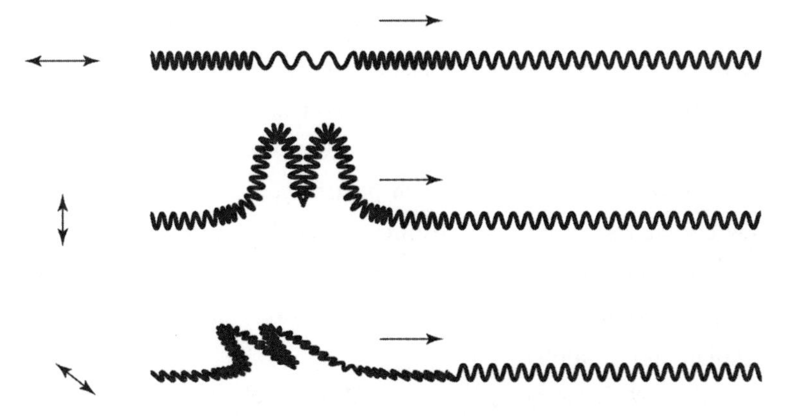

Figure 15. Creating longitudinal waves and two distinct transverse waves on a phone cord, respectively. (If anyone knows what a "phone cord" is anymore.)

The best way to visualize this is with a classic coiled phone cord, the type rarely used these days (fig. 15).[2] We imagine tying one end of this cord to a distant object and holding the other end taut. We first grab a handful of cord, pulling it toward us, and release it, resulting in a wave with alternating regions where the cord is stretched and compressed. This is a longitudinal wave, like a sound wave; the vibrations travel along the length of the cord, in the same direction as the wave is traveling. We may also, however, wiggle the free end up and down: the waves of the cord are also up and down, which is perpendicular to the left-to-right direction the wave itself is traveling. This is an example of a transverse wave. But we may also wiggle the cord side to side, producing waves that are also side to side; this is another transverse wave, distinct from the up-and-down one!

This is the explanation Young gave for the phenomenon of double refraction: light is a transverse wave, and for any direction light is traveling, there are two possible polarizations. Natural light, like sunlight, generally consists of a mixture of the two polarizations, which we call "up and down" and "side to side" for simplicity. The crystal itself has an interior structure that reacts differently to the two polarizations—the crystal is referred to as having an anisotropic structure. The word "anisotropic" comes from the Greek words *anisos* (unequal) and *tropikos* (turn). Because each polarization travels differently in the crystal, the result is two distinct refracted waves and two images. Young's explanation has been proven correct, and the transverse wave properties of light—polarization—are important in both the science of light and its application.

Though Young never became a full-time scientist again, his fundamental contributions to the theory of light were undeniable and became recognized worldwide. In the 1820s, he was elected to the American, French, Swedish, and Netherlands science academies.

True to his early polymath tendencies, Young spent his later years exploring a variety of subjects. He wrote several influential books on medicine and articles for *Encyclopedia Britannica* on numerous topics, including optics. He served on a number of public committees and commissions, including one to study the possible dangers of introducing gas lighting into London. He also exercised his talent in languages to work on deciphering Egyptian hieroglyphics, which were still cryptic at the time.

When Young died at the age of fifty-five, he was recognized as a singular genius of his time. A white marble tablet was erected at Westminster Abbey in 1834, with an epitaph by Hudson Gurney, the family friend whom Young had tutored so many years before. The epitaph reads:

> Sacred to the memory of Thomas Young, M.D., Fellow and Foreign Secretary of the Royal Society Member of the National Institute of France; a man alike eminent in almost every department of human learning. Patient of unintermitted labour, endowed with

> the faculty of intuitive perception, who, bringing an equal mastery to the most abstruse investigations of letters and of science, first established the undulatory theory of light, and first penetrated the obscurity which had veiled for ages the hieroglyphs of Egypt. Endeared to his friends by his domestic virtues, honoured by the World for his unrivalled acquirements, he died in the hopes of the Resurrection of the just.—Born at Milverton, in Somersetshire, 13 June 1773. Died in Park Square, London, 10 May 1829, in the 56th year of his age.

Young's discovery of the transverse wave nature of light would prove to be just as important to science as his law of interference because it answered many questions about the physics of light and raised new questions for future scientists to explain. And the phenomenon of double refraction and the Iceland spar that manifests it would also turn out to be important tools in bringing invisibility closer to reality.

7
Magnets and Currents and Light, Oh My!

This crystal tube the electric ray
  Shows optically clean,
No dust or haze within, but stay!
  All has not yet been seen.
What gleams are these of heavenly blue?
  What air-drawn form appearing,
What mystic fish, that, ghostlike, through
  The empty space is steering?

James Clerk Maxwell, "To the Chief Musician upon Nabla" (1874)

Like most major scientific discoveries, the recognition of the wave nature of light ended up raising as many questions as it answered. One immediate question that arose: What, in a light wave, is "waving"? In water waves, water does the waving, and in sound waves, the disturbance is carried in the density of air molecules. On a vibrating string, the string itself vibrates up and down, carrying the waves. But in Thomas Young's time, it was unclear what was being disturbed to produce a light wave. In analogy with the aforementioned examples, it was assumed that there was some sort of material permeating space that carried the vibrations of light, called the "aether" due to its mysterious and intangible quantities. But a better answer to the question

"What is waving in light?" would take some sixty years of research and speculation, leading to the recognition that light is something previously unimagined: a disturbance of electric and magnetic fields traveling through space, mutually maintaining each other. Today, we refer to light as an electromagnetic wave and recognize that infrared and ultraviolet radiation are just electromagnetic waves of different frequencies. The first clue to these discoveries would be uncovered in 1820 by a Danish philosopher, in an experiment that is utterly unique in the history of science.

For centuries, scientists and natural philosophers had considered electricity and magnetism to be two distinct phenomena in nature. Electricity was associated with small shocks on cold, dry days, lightning strikes, and faint attractive forces that could be generated by rubbing rods of amber with animal fur. Magnetism was associated with items like compasses and bar magnets. Electrically charged items seemed to attract or repel each other completely; magnetic items possessed North and South poles, where like poles would repel, and unlike poles attract.

There were a number of isolated experiments that hinted at a link between electricity and magnetism: Benjamin Franklin had magnetized needles by discharging electricity through them, and sailors had occasionally reported a reversal of the polarity of compass needles after their ships were struck by lightning. Also, it was lost on nobody that both electricity and magnetism involve similar attractive and repulsive forces. Nevertheless, other experiments were inconclusive, and most prominent scientists dismissed a significant connection. In his book on *Natural Philosophy* (1807), Thomas Young wrote, "There is no reason to imagine any immediate connexion between magnetism and electricity, except that electricity affects the conducting powers of iron or steel for magnetism, in the same manner as heat or agitation."[1]

Into this state of affairs came the Danish philosopher Hans Christian Oersted (1777–1851). The son of a pharmacist, Oersted gained an early interest in science by working in his father's shop. His early education came from self-study at home, which served him quite well: he entered the University of Copenhagen in 1793 and excelled there, earning honors for papers on aesthetics and physics. He earned his doctorate in 1799, at age twenty-two, for a dissertation on the works of philosopher Immanuel Kant entitled "The Architectonicks of Natural Metaphysics."

Oersted's philosophical training would play a crucial role in his discovery of electromagnetism. In 1801 he received a scholarship that allowed him to spend several years traveling through Europe. While in Germany, he immersed himself in the country's philosophical circles, where he became acquainted with the Naturphilosophie movement discussed earlier. He even spent time with Johann Wilhelm Ritter, the discoverer of ultraviolet light and strong proponent of Naturphilosophie, and developed an interest in how the philosophical movement could be used to make new discoveries in science. Followers of Naturphilosophie, who believe that all phenomena in nature are connected, naturally believed in the possibility that electricity and magnetism themselves were somehow related to each other.

Oersted became a professor at the University of Copenhagen in 1806 and brought with him the convictions borne of Ritter and Naturphilosophie. Over the next few years, he did research on electricity and acoustics and at the same time grew the physics and chemistry programs of the university.

Over the winter term of 1819–20, Oersted was giving a series of lectures on electricity and magnetism. On April 21, 1820, he was slated to give a lecture on the well-known similarities between electrical and magnetic forces, and he thought back to his own beliefs regarding their connection. He wondered again whether it is possible to produce any sort of magnetic effect using a flowing current of electricity and decided to prepare a small experiment to look for such an effect. In this case, he placed a length of wire over a compass, with glass between them. The idea was to see whether there was any effect on the compass when an electrical current ran through the wire.[2]

We turn to Oersted's own words for what happened next, written in the third person, some ten years afterward,

> The preparations for the experiments were made, but some accident having hindered him from trying it before the lecture, he intended to defer it to another opportunity; yet during the lecture, the probability of its success appeared stronger, so that he made the first experiment in the presence of the audience. The magnetical needle, though included in a box, was disturbed; but the effect was very feeble, and must, before its law was discovered, seem very irregular, the experiment made no strong impression on the audience.[3]

With a single twitch of a compass needle, Oersted had demonstrated that electrical currents will produce a magnetic effect, and in doing so he revolutionized physics. Because he did not have a chance to test his experiment beforehand, this demonstration was done in a lecture hall, with students watching. It is possibly the only major scientific discovery in history to have been made for the first time before a live audience. It is perhaps not surprising that the experiment "made no strong impression," considering the audience didn't know what they were looking at.

Oersted did not immediately announce his results to the world, and it took several months for him to do further experiments. By July 1820, however, he had done enough testing and drafted a paper detailing his observations. The paper appeared first in Latin but was quickly translated into a multitude of languages, including English:

"Experiments on the Effect of a Current of Electricity on the Magnetic Needle."[4] Oersted's work was met with instant acclaim, and scientists throughout Europe traveled to see him to discuss his work.

With the discovery that electricity could produce magnetic effects, researchers naturally began to wonder if the opposite is true: Could a magnet produce electrical effects? Over the next decade, many researchers tried and failed to make such a connection. In the end, it was the British scientist Michael Faraday, a man who started his career as a humble apprentice bookbinder, who would provide the next groundbreaking discovery.

Born the son of a blacksmith in 1791, Michael Faraday lived when science was almost exclusively the provenance of the upper class. He was raised with little formal education, and when he became an apprentice bookbinder at age fourteen, he seemed destined for an unexceptional life. Faraday was a voracious reader, however, and his job at George Riebau's bookshop gave him access to all the books he could wish for. The first science book that captured his attention was an encyclopedia volume from 1797 containing an article on electricity. Intrigued, he spent some of his meager savings on his own electrical apparatus and tried to reproduce some of the results discussed in the article.

In 1810, another book arrived at Riebau's: it was *Conversations on Chemistry*, by Jane Marcet. Self-educated in chemistry, Marcet performed experiments at home with the help of her husband, Alexander, a physician. Wishing to learn more, she attended the lectures of the famed chemist Humphry Davy at the Royal Institution, the same institution where Thomas Young had worked, and this set in her a determination to write a popular book on chemistry that would be understandable for women like her who had little exposure to science otherwise. The result, *Conversations on Chemistry*, was published in 1805 and later went through sixteen editions.

Marcet's work was also ideal reading for the self-taught Faraday. He devoured its contents, learned from it, and was inspired to pursue chemical studies. By 1812, as his bookbinding apprenticeship was coming to an end, he was eager to get himself a scientific job, if at all possible. In addition to his love of scientific study, Faraday viewed the trade professions as "vicious and selfish," and viewed science as making its pursuers "amiable and liberal," a noble cause he aspired to.[5]

Here a combination of determination and fate would set Faraday on his scientific course. Through his job, Faraday had become friends with William Dance, a distinguished pianist and violinist, and Dance was impressed with Faraday's diligence and hard work. In early 1812, Dance secured tickets for Faraday to Humphry Davy's chemistry lectures, and the young man took detailed notes of everything he learned. Further encouraged by Dance, Faraday wrote to Davy to inquire about job prospects at the Royal Institution, and he included his lecture notes for good measure. In late December 1812, Davy replied graciously, agreeing to meet the young man as soon as possible.

Faraday, now working as a journeyman bookbinder, had five weeks of anxious waiting before Davy returned to town and they could meet. The result was worth the wait, however: Davy informed Faraday that a job as a chemical assistant had opened up in the Royal Institution and he felt that Faraday would be good for the job. Somewhat sardonically, however, Davy suggested that bookbinding might be a more stable and noble profession to stay in, as Faraday later recalled:

> At the same time that he thus gratified my desires as to scientific employment, he still advised me not to give up the prospects I had before me, telling me that Science was a harsh mistress, and in a pecuniary point of view but poorly rewarding those who devoted themselves to her service. He smiled at my notion of the superior

> moral feelings of philosophic men, and said he would leave me to the experience of a few years to set me right on that matter.[6]

The new job did not start right away, so Faraday continued his work as a bookbinder for another month, while also occasionally helping Humphry Davy as a note taker. Davy had injured his eyes in a chemical explosion late the previous year, and his vision had not fully recovered by February 1813. In March, however, Faraday was announced as the new chemical assistant at the Royal Institution.

As great as this opportunity was, Faraday would soon get an even greater one. Not long after Faraday's appointment, Davy resigned his formal professorship at the Royal Institution and made plans to begin a multiyear tour of continental Europe later that year. With war brewing, however, Davy's valet refused to take the potentially dangerous journey. Davy immediately offered to bring Faraday along on the trip in the dual role of valet and chemical assistant.

The opportunity to meet some of Europe's most distinguished scientists and see the continent was too good to pass up, despite the ignominy of working as a servant. The trip was largely rewarding: Faraday impressed Davy's peers with his skills and his thoughtfulness, and the contacts would serve him well in the future. He even had the opportunity to meet Jane Marcet in Geneva, where she and her husband were summering. Lady Davy, however, viewed Faraday more as a servant than a scientist and insisted that he take dinner with the servants. After the ladies retired to the parlor, Marcet's husband, Alexander, whispered, "And now, my dear Sirs, let us go and join Mr. Faraday in the kitchen."[7]

On returning to London, Faraday resumed his role as chemical assistant at the Royal Institution. He soon became recognized as an excellent public lecturer as well as a brilliant experimenter, and his fame grew. In 1825, he became director of the Laboratory of the Royal

Institution, and in 1833, he was appointed the Fullerian Professor of Chemistry at the Royal Institution, a position created specifically for him, which he held for the rest of his life.

In the 1820s, Faraday began, in earnest, to investigate electricity, a subject that had fascinated him since his bookbinding days. He learned of the works of Oersted and, expanding on them, in 1821 built the world's first electric motor, a device now known as a homopolar motor. Over the next decade, Faraday would perform experiments in optics, electricity, and chemistry. In 1831, Faraday started releasing a long-running series of papers to the Royal Society titled "Experimental Researches in Electricity." Almost immediately, he announced his most famous discovery, one that would eventually transform the world, which we now know as electromagnetic induction.[8]

Faraday had been occupied with the question that had obsessed researchers ever since Oersted's revelation in 1820: If an electric current can produce magnetism, is it possible for a magnet to somehow produce electricity? The thinking of Faraday and others was of a sort of reciprocal arrangement: if an electric current can somehow attract a magnet, then a magnet must be able to distort an electric current.

In Faraday's crucial experiment, he wound an extremely long wire around a cylinder of wood, connected to a battery with a switch (fig. 16). It was known by this time that a current running through such a coiled wire could produce a very strong magnetic field, which increases with the number of coils. On top of this structure, but electrically insulated, Faraday wound another long wire in a coil, and this one was connected to a meter that measures electrical current (known as a galvanometer at the time, in honor of electrical pioneer Luigi Galvani).

With the current on, Faraday saw no signal on the galvanometer. However, he noticed something strange: the needle of the galvanometer twitched the instant he turned the current on or off. It was a

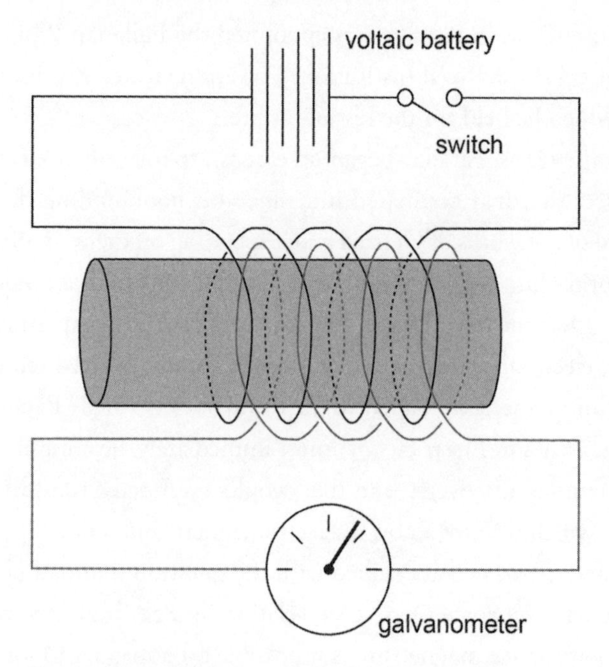

Figure 16. Faraday's experiment for electromagnetic induction.

small effect, like the twitch of the compass needle in Oersted's original experiment, and just like Oersted's experiment, it represented a profound discovery.

With more experiments, Faraday concluded—correctly—that a *change* in magnetism will induce an electrical current in a wire. This showed that the connection between electricity and magnetism worked both ways, though apparently differently in each direction. Faraday's result was a revelation to the scientific community, which embraced the results.

Faraday made many other important discoveries, such as demonstrating that different manifestations of electricity—from chemical reactions, from friction, and from animals—are all equivalent. He even tested the shocking powers of creatures like electric eels by putting his hands on them to estimate the amount of shock they could give. At the age of fifty-four, he made the observation that a magnetic field can change the polarization of light, hinting at a connection between light, electricity, and magnetism. The final piece of the puzzle connecting all three, however, would elude him.

Jane Marcet and Michael Faraday remained friends for their entire lives and kept regular correspondence. Marcet would write to Faraday for explanations of his new discoveries, to be included in the latest edition of *Conversations on Chemistry.* Faraday, for his part, did not hesitate to acknowledge the importance of her work on his own life:

> Do not suppose that I was a very deep thinker, or was marked as a precocious person. I was a very lively, imaginative person, and could believe in the "Arabian Nights" as easily as in the "Encyclopaedia"; but facts were important to me, and saved me. I could trust a fact, and always cross-examined an assertion. So when I questioned Mrs. Marcet's book by such little experiments as I could find means to perform, and found it true to the facts as I could understand them, I felt that I had got hold of an anchor in chemical knowledge, and clung fast to it.[9]

In 1852, at sixty-one years of age, Faraday introduced what would be his greatest *theoretical* accomplishment, and it would turn out to be just as important as his experimental contributions, or even more so.

Since the time of Newton, forces between two objects, such as gravity, electricity, and magnetism, had been mathematically viewed as a direct interaction between those two objects, often referred to as "action at a distance." What caused this action at a distance was un-

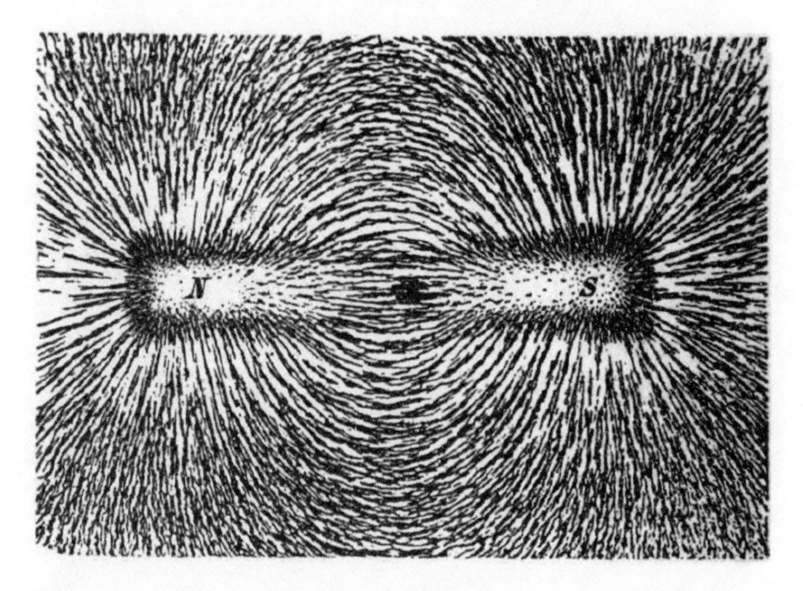

Figure 17. Iron filings arranged around a bar magnet. Image from Newton Henry Black and Harvey N. Davis, *Practical Physics* (New York: Macmillan, 1913), 242, fig. 200.

clear and much debated. Some viewed magnetism as an otherwise unobservable fluid, emanating from the North pole of a magnet and circulating into the South pole.

This view of magnetism as a fluid was strengthened by an experimental demonstration that is still used today, albeit with a very different interpretation. In the demonstration, a magnet is placed under a piece of paper, and iron filings are sprinkled on the paper above it. The result is that the filings turn in place, clump together, and form what appear to be lines of metal, stretching from the North pole of the magnet to the South pole (fig. 17).

To many researchers, this suggested that the filings were being oriented and clumped together by the flow of the mysterious magnetic liquid. Faraday, however, saw something much more practical in the distribution of metal shavings: a means of quantifying the direction and strength of the magnetic force around a magnet. In the twenty-eighth paper of his series on electricity and magnetism, he introduced his idea of magnetic lines of force as follows: "A line of magnetic force may be defined as that line which is described by a very small magnetic needle, when it is so moved in either direction correspondent to its length, that the needle is constantly a tangent to the line of motion."[10] In other words, he imagined that a magnet is surrounded by lines of force that can be detected at any location by placing a single iron filing in that location. The entire collection of lines he referred to as a magnetic field. An electric field for a collection of electric charges can also be defined using a single pointlike electric charge, called a test charge; the field lines point in the direction that the test charge is pushed or pulled when placed in the vicinity of the collection.

Today, physicists can readily calculate what the fields look like for simple systems. A bar magnet has lines of force circulating from the North pole to the South pole, and back around again through the magnet. A long wire carrying electric current produces a magnetic field, as Oersted showed, that circulates around the wire. A pointlike electric charge produces lines of force that point directly away from it if the charge is positive, and towards it if negative.

Drawings of fields only show a handful of field lines out of what are in principle an infinite number of lines filling all of space. These pictures are intended to give a researcher an intuition for the behavior of magnets and electrical charges. But Faraday also noted that there is a quantitative implication to these pictures: the strength of the force is directly proportional to the density of field lines at that point in space. As one goes farther from the sources of the fields, the lines become less dense and the forces correspondingly less intense.

Faraday was very careful to argue that he intended his idea of lines of force to be a conceptual tool and not a physical explanation of electricity and magnetism. In this paper, he said, "I desire to restrict the meaning of the term *line of force,* so that it shall imply no more than the condition of the force in any given place, as to strength and direction; and not to include (at present) any idea of the nature of the physical cause of the phenomena; or to be tied up with, or in any way dependent on, such an idea."[11] But Faraday's construction resulted in one very important, even crucial, philosophical change in the way researchers viewed such forces: it naturally replaced the idea of action at a distance with an intermediate cause for the attraction or repulsion of objects. It became reasonable to view, for example, a magnet as producing a magnetic field, and that magnetic field in turn exerting a pull on a different magnet. Instead of saying, "Magnet A attracts magnet B, and vice versa," it became reasonable to say, "Magnet A produces field A, which interacts with magnet B, causing an attraction, and vice versa." Action at a distance was replaced with the idea that objects interact through the fields that they produce. This idea of fields is now a key part of our description of all fundamental forces in nature.

With Faraday's description, phenomena that were described in somewhat awkward terms with existing language can be described much more elegantly. Faraday induction, for example, can be described as "a magnetic field changing in time induces a circulating electric field." If a wire loop is placed in that circulating electric field, then the electric field drives an electrical current in the wire.

Faraday's description of lines of force and fields did not gain much immediate attention. Faraday himself was not trained in mathematics, and his descriptions failed to convince theoretical physicists at the time of the usefulness of his concepts. It would be almost a decade before the brilliant Scottish scientist James Clerk Maxwell recognized the power and utility of Faraday's ideas.

James Clerk Maxwell was born in Edinburgh, Scotland, in 1831 into a family of comfortable means. Like Young, he displayed his remarkable intelligence and curiosity at a very young age. By age three, he would question the workings of everything, asking, "What's the go o'that? What does it do?" and would follow up with, "But what's the *particular* go of it?" if the initial answer was too vague. As a hint of his future life's path, young Maxwell was also fascinated by colors and their origins; once when told that a stone was the color blue, he asked, "But how d'ye know it's blue?"[12]

Maxwell's mother, Frances, provided his early education until she suffered an untimely death when he was only eight years old. His father and aunt-in-law took over the remainder of this early tutoring. In February 1842, his father took him to a public demonstration of electromagnetic machines including a small electric-powered train and an electric-powered saw; this demonstration likely had a strong influence on young Maxwell's future interests.[13]

Maxwell's career and talents progressed at an astonishing speed. He wrote his first scientific paper, "On the Description of Oval Curves and Those Having a Plurality of Foci," at age fourteen.[14] It was presented to the Royal Society of Edinburgh by Professor James Forbes of the University of Edinburgh; Maxwell was considered too young to present it himself. At sixteen, Maxwell began taking classes at the University of Edinburgh. At nineteen, he transferred to Cambridge, where he continued his studies at Trinity College, earning a degree in mathematics in 1854. While at Trinity, Maxwell studied the perception of colors, continuing the fascination that had begun when he was a child. In 1855, he presented a paper to the Cambridge Philosophical Society, "Experiments on Colour," describing the principles of color combination; for the first time, he was able to present the paper himself.[15] By late 1855, Maxwell had accepted a fellowship to teach at Trinity, and by late 1856, he had accepted a professorship in natural philosophy at Marischal College, Aberdeen, at only twenty-five years old.

Incredibly, Maxwell would find himself laid off in only a few years, when the University of Aberdeen merged with King's College of Aberdeen in 1860. He moved to London with his wife, Katherine, where he took a professorship at King's College of London. It was in London that he would do some of his most groundbreaking research, including the creation of one of the earliest methods for taking color photographs. He also attended lectures by Michael Faraday, who, now in his seventies, was still working at the Royal Institution.

Maxwell had taken an early interest in Faraday's ideas of "lines of force" and had written a paper in 1855 strongly encouraging others to pay closer attention to them.[16] Researchers had treated Faraday's work as a nonrigorous description of electric and magnetic forces, merely a visual aid, but Maxwell argued that the concept of lines of force and fields of such lines could be made mathematically rigorous.

Maxwell's most magnificent breakthroughs were made in a four-part paper, "On Physical Lines of Force," published in 1861. In the first part, Maxwell took a bold stance, suggesting that the lines of force introduced by Faraday are not a mere mathematical convenience but a real phenomenon. Again, the magnet and iron filings experiment was a strong justification: "The beautiful illustration of the presence of magnetic force afforded by this experiment, naturally tends to make us think of the lines of force as something real, and as indicating something more than the mere resultant of two forces, whose seat of action is at a distance, and which do not exist there at all until a magnet is placed in that part of the field."[17] More important, Maxwell recognized that something was missing in the existing descriptions of electric and magnetic phenomena, considering that Oersted and Faraday had already shown that electricity and magnetism are related. We may reason Maxwell's idea as follows: If a time-varying magnetic field produces a circulating electric field, as Faraday had shown, shouldn't a time-varying electric field produce a circulating *magnetic field?* Maxwell noted that a time-varying electric field would act very much like an electric current, and he called such time-varying fields displacement currents: they produce magnetism just like an ordinary current, as Oersted had shown, but without any electric charges being present.

The most important result of Maxwell's paper, however, came from treating electricity and magnetism as disturbances of a hypothetical material medium, which was referred to as the aether. By analyzing the speed at which waves would travel in this medium, Maxwell found that the speed was remarkably close to the known speed of light. He then concluded, "We can scarcely avoid the inference that *light consists in the transverse undulations of the same medium which is the cause of electric and magnetic phenomena.*"[18] In other words: what we perceive as light is, in fact, a transverse wave consisting of oscillating electric and magnetic fields. Maxwell had officially hypothesized the unity of electricity, magnetism, and light as a single phenomenon that we now call electromagnetism.

Only a few years later, in 1865, Maxwell published an improved mathematical description of all electric and magnetic phenomena together and showed that these equations, when the displacement current is included, explicitly result in a mathematical wave equation that predicts waves traveling at the speed of light.[19] This discovery marked the beginning of modern optical science, and the equations that Maxwell introduced are known as Maxwell's equations. These equations are still used today by modern optical researchers, with changed notation but the same content.

Today, using Faraday's field lines and Maxwell's mathematics, we

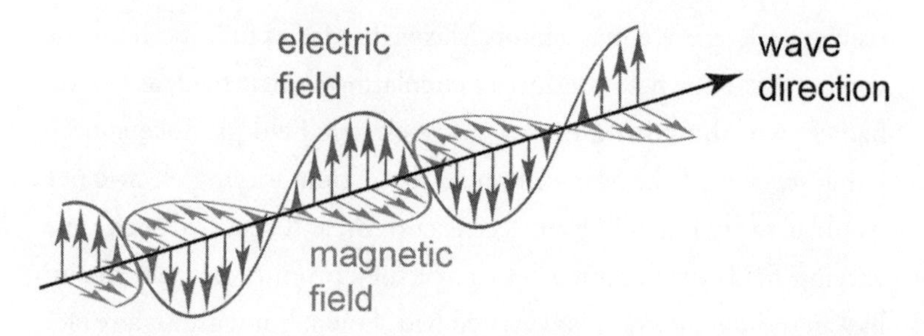

Figure 18. An electromagnetic wave. The electric field and magnetic field are at right angles to each other, and both oscillate transverse to the direction the wave is traveling. It is a transverse wave.

have a clear idea of what an electromagnetic wave "looks" like. It consists of oscillating electric and magnetic fields, perpendicular to each other and both perpendicular to the direction the wave is going, moving at the speed of light (fig. 18).

How does an electromagnetic wave form from the distinct phenomena of electricity and magnetism? One can roughly understand it from Faraday induction and Maxwell's displacement current. Faraday's law says that a changing magnetic field creates an electric field, and Maxwell's displacement current indicates that a changing electric field produces a magnetic field. Therefore, once created, these magnetic and electric fields keep themselves oscillating as they evolve in time, producing a wave that can travel long distances. The process can be started by an antenna generating an oscillating electrical current, which produces an oscillating magnetic field, as Oersted had shown. This is the mechanism by which radio antennas, and cellphones, produce broadcast signals.

Though Faraday had been no mathematician, Maxwell gave him tremendous credit in the development of his own theoretical discoveries. In his *Treatise on Electricity and Magnetism* (1873), he wrote fondly of Faraday and his achievements.

> As I proceeded with the study of Faraday, I perceived that his method of conceiving the phenomena was also a mathematical one, though not exhibited in the conventional form of mathematical symbols. I also found that these methods were capable of being expressed in the ordinary mathematical forms, and thus compared with those of the professed mathematicians.
>
> For instance, Faraday, in his mind's eye, saw lines of force traversing all space where the mathematicians saw centres of force attracting at a distance: Faraday saw a medium where they saw nothing but distance: Faraday sought the seat of the phenomena in real actions going on in the medium, they were satisfied that they had found it in a power of action at a distance impressed on the electric fluids.[20]

Maxwell had theoretically shown that electromagnetic waves exist and that they travel at the speed of light, but experimental verification was needed. The key experiments were done by the German physicist Heinrich Hertz between 1886 and 1889, using what we would now recognize as a crude radio wave transmitter and receiver.[21] He used a mirror to reflect radio waves back to their source, creating standing waves, like the sound waves produced inside of an organ pipe. By measuring the wavelength of the waves—roughly 9.3 meters—and their frequency, he could calculate the speed of the waves and confirm that the speed was, within experimental error, equal to the speed of light. Hertz's demonstration therefore showed that electromagnetic waves exist and that they traveled at the speed of light; after this, few researchers doubted that light itself was an electromagnetic wave.

In an often-told story, Hertz was asked by one of his students about what his discovery could be used for, and he reportedly replied, "Nothing I guess."[22] This story, whether true or not, is particularly amusing because of how wrong the prediction would turn out to be.

In only a few short years, Guglielmo Marconi and Nikola Tesla began developing radio communications, ushering in a new era of technology and, consequently, tremendous changes to society.

From a scientific perspective, the discovery and confirmation of electromagnetic waves resulted in a greater understanding of light and what it can and cannot do. For the science of invisibility, it would in the short term result in authors of fiction imagining more refined ideas of how invisibility could work, ideas that would not conflict with this new understanding. Fortunately, one author in particular was up to the challenge, and he would write some of the greatest science fiction stories of all time.

8

Waves and Wells

In the next experiment, and in those succeeding, he allowed his chemical agents to take firmer hold upon the tissues of my body. I became not only white, like a bleached man, but slightly translucent, like a porcelain figure. Then again he paused for a while, giving me back my color and allowing me to go forth into the world. Two months later I was more than translucent. You have seen floating those sea radiates, the medusa or jellyfish, their outlines almost invisible to the eye. Well, I became in the air like a jellyfish in the water.

Edward Page Mitchell, "The Crystal Man" (1881)

Though the nineteenth century had been filled with revelations about the nature of light, electricity, and magnetism, it saved one last surprise to be revealed in its closing years: a new, mysterious form of radiation, unseen by the naked eye, that could penetrate through all but the densest materials as if they were invisible. These rays, initially named after their discoverer, much to his annoyance, as Röntgen rays, eventually became known by his preferred name that captured their mysterious nature: X-rays. These X-rays would not only revolutionize science and medicine but would inspire a science writer named H. G. Wells to write the most famous story of scientific invisibility, *The Invisible Man.*

The momentous discovery of X-rays was only possible because of a discovery, decades earlier, of another sort of mysterious ray. The seeds of this earlier discovery can be traced back, perhaps unsurprisingly, to Michael Faraday. In 1838, Faraday was investigating the discharge of electric sparks in a gap between metal surfaces, using an electrical device that produced a negative charge on one surface (later called the cathode) and a positive charge on the other (later called the anode). He was particularly interested in seeing how the behavior of the spark changed when it was generated in different gases, and he placed the entire apparatus in a glass jar that he could fill with gas or, alternatively, create a partial vacuum within. He found that under certain circumstances, no spark would be created but ghostly glowing columns would be formed in the jar, extending from the cathode and anode toward each other. Faraday was not in a position to explain this phenomenon in his time, but it would later be recognized as arising from electrons colliding with atoms as they flow from the cathode to the anode. In the collisions, electrons transfer energy to the atoms; that energy is then reemitted as light.

In 1857, the German glassblower Heinrich Geissler was able to generate a better vacuum in a tube, with a lower density of gas inside, and he found that he could make the entire interior of the tube glow using electricity. (This technique is basically how glowing neon signs work today; they are of course filled with neon gas.) Geissler's employer, Julius Plücker, worked with Geissler's tubes over the next year and found that, by evacuating even more gas from the tube, the wall of the tube opposite to the cathode could be made to glow. Plücker brought a magnet near the glow of the tube wall and found that he could distort the shape of that glow. Because electrical currents have their motion affected by a magnetic field, Plücker's observation suggested that the glow of the tube wall was caused by some sort of moving electrical disturbance.

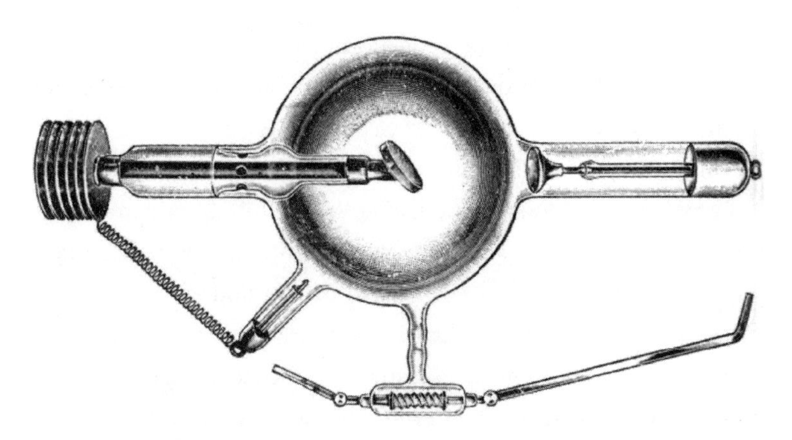

Figure 19. A Cossor tube, a later-model cathode ray tube used to produce X-rays. The cathode is on the right; the center of the tube is the "anti-cathode," which replaced the anode in X-ray tubes. A secondary anode, extending to the lower left, was included to stabilize the tube operation. Illustration from Kaye, *X Rays* (1918).

In 1869, Johann Hittorf, a pupil of Plücker, improved the vacuum inside a Geissler tube further and demonstrated that objects placed within the tube caused a shadow to be cast in the glowing wall. This suggested to researchers that the electricity was being carried by some sort of rays, as light was once thought to be, that travel in straight lines from the cathode to the anode end of the tube. Because of this, these mysterious rays were called cathode rays (fig. 19).

History then repeated itself, as scientists debated: Are cathode rays a wave or a particle? French and English scientists tended to believe that they must be particles, while German scientists believed them to be waves. Cathode rays would be intensely studied and furiously debated for the rest of the century. In 1897, the debate was ended through the work of the English physicist J. J. Thomson. Using electric fields and magnets, Thomson was able to determine the ratio of electric charge to mass for the mysterious cathode rays, showing that they consist of a stream of identical particles of negative charge and low mass. This was the first fundamental building block of nature discovered, and it was named the electron.

Two years before Thomson's momentous discovery, a fifty-year-old professor of physics at the University of Würzburg named Wilhelm Conrad Röntgen was studying the properties of cathode rays. Researchers had already demonstrated that cathode rays could pass through an aluminum "window" put on a cathode ray tube; if these rays were allowed to impact a nearby screen coated with fluorescent paint, they would make the screen glow. Röntgen was interested in determining whether the cathode rays could also pass through the glass of a tube that possessed no window.

In the process of preparing his experiments, he wrapped his cathode ray tube in cardboard to make it light-tight; he wanted to ensure that any glow he detected came from the fluorescence of the screen and not the familiar glow of the tube itself. On November 8, 1895, while testing his wrapped tube in a darkened room, he noticed that a fluorescent screen some distance away was glowing. This could not be the cathode rays themselves, which had already been shown to travel only a few centimeters through air. Some other type of mysterious ray was not only passing through the glass of the tube and the cardboard wrapping but traveling long distances in air and still exciting fluorescence. Röntgen dubbed these mysterious new rays "X-rays."

The amazing ability of these rays was immediately apparent, as they could seemingly go through just about any substance. Röntgen quickly realized the potential of such rays for imaging the interior of the human body, and in his first published paper he included an X-ray image of his wife's hand, taken on December 22, 1895 (fig. 20).[1] When she saw the image, she reportedly said, "I have seen my death!"

The medical significance of Röntgen's work was immediately ob-

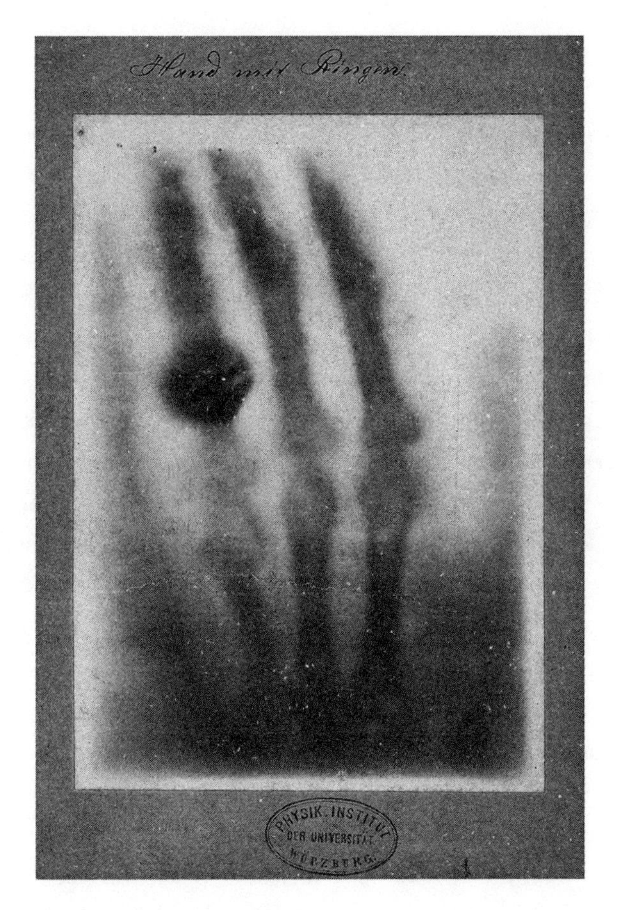

Figure 20. "I have seen my death!" A print of Röntgen's X-ray of his wife's hand.

vious. His scientific paper was published on December 28, 1895, and the first news account of his discovery was published in the *Presse* of Vienna on January 8, 1896. From the account of the *Presse,* "The surgeon could then determine the extent of a complicated bone fracture without the manual examination which is so painful to the patient:

he could find the position of a foreign body such as a bullet or a piece of shell much more easily than has been possible heretofore and without any painful examinations with a probe."[2]

But what were these new X-rays? Initial investigations showed that they had a number of baffling properties. They did not appear to reflect or refract the way that ordinary light or other invisible rays like infrared and ultraviolet rays did. Eventually, however, they were shown to be yet another type of electromagnetic wave, one with a wavelength much shorter than that of visible light. Where visible light has a wavelength on the order of 500 nanometers (billionths of a meter), the *longest* wavelength X-ray has a wavelength of 10 nanometers. The energy of X-rays is so high that they go through most objects as if the objects are not even there, leading to their incredible imaging properties.

How are X-rays created? It would later be found that they are generated by an extreme version of the same process that produces all electromagnetic radiation—acceleration of electric charges. This is a consequence of Maxwell's equations: when an electric charge accelerates, it produces a time-varying electric and magnetic field, resulting in an electromagnetic wave. X-rays are generated by extreme accelerations of electrons in a cathode ray tube. When the electron leaves the cathode, its speed increases as it approaches the positively charged anode. In a vacuum tube, there is no gas present to slow down the electrons, which then collide with the anode at tremendous speeds.[3] This collision is a rapid deceleration of the electron, producing a high-energy X-ray.

At the time of their discovery, Röntgen's X-rays became an international sensation, both within scientific circles and among the general public. Misinformation spread almost as quickly as news of the discovery itself: If X-rays can see through anything, might people be able to use them to spy on their neighbors and see through their

Figure 21. "An enterprising man from London has even advertised X-ray proof underclothing." Illustration from the Bridgewater, N.J., *Courier-News,* May 27, 1896. Somehow, the X-ray manages to image the man with his right hand pointed upward, even though he is holding it against his waist.

clothing? One newspaper article debunking such assertions noted that "an enterprising man in London has even advertised X-ray underclothing" (fig. 21).[4]

It is not difficult to see how X-rays immediately became associated with invisibility among the public. The rays themselves were invisible, and they could effectively make other things, such as the human body, seem invisible in turn. This possibility caught the imagination of a science journalist named H. G. Wells, who would take his own scientific knowledge, mixed with popular conceptions of X-rays, and turn them into one of the most famous science fiction stories of all time.

Herbert George Wells was born in 1866 in Bromley, in southeast London. His family struggled financially—his mother worked as a shopkeeper and his father earned income irregularly as a professional cricket player. Wells struggled through much of his young adult life and even went hungry at times; these experiences would motivate his later efforts to promote a fair and even utopian society.

Wells's life and future career were ironically transformed for the better by misfortune. In 1874, when he was seven or eight years old, he broke his leg while roughhousing with an older boy. While laid up and recovering, he took to reading and found that he had a real love of books. A steady supply of new material was brought to him during his recovery by his father and the guilt-ridden mother of the boy who had inadvertently injured him. Wells credited that incident for his future career, later saying, "Probably I am alive today and writing this autobiography instead of being a worn-out, dismissed and already dead shop assistant, because my leg was broken."[5]

His early education was fraught with struggle and disappointment. After recovering from his leg injury, he entered a private academy run by Thomas Morley. Morley was an uninspiring teacher; he focused on rote memorization, had irregular moods and was bad tempered with the students, and often did not understand the subjects he was teaching. Wells later wryly recalled, "It is very difficult to give any facts about this dominie and his Academy which do not carry with them a quality of Dickens-like caricature."[6] In 1877, Wells's father suffered a serious broken leg, and that put an end to his cricket career—and the family income. Wells was moved through several apprenticeships, twice as a draper and once as a chemist's assistant. He regarded his time as a draper as the most miserable of his life. Finally, in 1883, he convinced his parents to release him from his draper commitments—even taking the extreme measure of threatening suicide if he did not get what he wanted—and found work as a pupil-teacher, a student who aids in the teaching of younger students, at Midhurst Grammar School.

Working at Midhurst finally gave Wells the opportunity to study subjects in the depths that he desired, and he submerged himself in every scientific topic imaginable. One of his teachers, Horace Byatt, even prepared bogus classes for Wells, basically study halls, that were simply opportunities for Wells to read whatever he liked. During those sessions, Byatt worked on his personal correspondence. The hard work paid off for Wells: his examination grades were so high that he earned a scholarship at the Normal School of Science (later the Royal College of Science) after only a year at Midhurst. There, he studied biology under Thomas Henry Huxley, a man famously known as "Darwin's bulldog" for his fierce and public defense of Charles Darwin's theory of evolution. Wells had a less satisfying education in physics, and he felt that his teacher perhaps did not understand the subject himself. In hindsight, Wells felt that this was, if anything, foreshadowing: "I did not realize it then, but at that time the science of physics was in a state of confusion and reconstruction, and lucid expositions of the new ideas for the student and the general reader did not exist."[7]

In 1887, he left the Normal School and found work as a teacher in Wales at the Holt Academy. Though Wells was optimistic that this would be an acceptable location, on arrival at the academy he found it a dismal, uninspiring place and felt that he had steered his career into a "cul-de-sac." But again misfortune would eventually change his fate for the better. While playing football, he was fouled hard by another player. Wells found himself immediately ill and bedridden; doctors concluded that the impact had crushed one of his kidneys. Even more ominous, some days later Wells began coughing up blood, which doctors interpreted as the deadly disease tuberculosis. But this diagnosis allowed Wells to negotiate a leave of absence from work with pay, and he returned home to stay with his mother and await his fate.

A doctor named Collins was staying in the same lodging house and undertook to care for Wells. He refused to accept a diagnosis of tuberculosis and diligently treated Wells, who slowly began to recover. While convalescing, Wells started writing prolifically, tackling every style of writing he knew of: short stories, short essays, poetry. After several months, on reviewing his early efforts, he burned them all and started again, working to improve his style. All the while his condition improved, albeit slowly. Finally, in the summer of 1888, he came to a significant decision:

> One bright afternoon I went out by myself to a little patch of surviving woodland amidst the industrialized country, called "Trury Woods." There had been a great outbreak of wild hyacinths that year and I lay down among them to think. It was one of those sun-drenched afternoons that are turgid with vitality. Those hyacinths in their upright multitude were braver than an army with banners and more inspiring than trumpets.
>
> "I have been dying for nearly two-thirds of a year," I said, "and I have died enough."
>
> I stopped dying then and there, and in spite of moments of some provocation I have never died since.[8]

Wells returned to London in search of a career in writing, but it did not come easily. To pay the bills he again took a job as a teacher and even worked to earn a bachelor of zoology degree in 1890 from the University of London External Programme. A relapse of illness in late 1890 gave Wells time to try writing science for a popular audience, and the results were mixed. After his first article, "The Rediscovery of the Unique," was published in the *Fortnightly Review,* he submitted another article to the magazine titled "The Universe Rigid," which attempted to describe the principles of four-dimensional space-time in nontechnical terms. The editor W. E. Henley, on reading the article proofs, was so incensed by its incomprehensibility that he summoned Wells personally to his office to yell at him: "He caught up a proof beside him and tossed it across the table. 'Dear Gahd! I can't understand six words of it. What do you mean by it? For Gahd's sake tell me what it is all about? What's the sense of it? What are you trying to say?'"[9]

Wells honed his writing skills and aimed for less mind-boggling topics, and by 1894 he was writing regular articles for the *Pall Mall Gazette,* which earned him more money than he had ever made teaching. In 1895, the *Gazette* was looking for longer serials to be published, and Wells proposed a series of articles about the possibility of time travel, something he had pondered and written about in a short story for his college newspaper under the title "The Chronic Argonauts." His story was serialized in the *New Review* from January to May 1895 under the now more familiar and famous name *The Time Machine.* At that point, the editor of the *New Review* was W. E. Henley, the same man who had blasted Wells's incomprehensible science writing years before. Despite Henley's earlier frustration, something positive about Wells clearly had stuck with him. Ironically, *The Time Machine* introduces the premise of moving through the "fourth dimension" of time, the very concept that had confused and enraged Henley before.

The Time Machine was a great success and published in its first book edition in May 1895. From there, Wells's career as a novelist was assured. His next novel of science fiction, published in 1896, was *The Island of Doctor Moreau,* about a scientist creating humanlike hybrid animals and the horrific consequences of such actions. But Wells's third novel, published in 1897, would be perhaps his most famous: *The Invisible Man* (fig. 22).

H. G. WELLS'S NEW ROMANCE.

THE INVISIBLE MAN

Other writers have treated this theme, but they have generally given the invisible man a power which it was something more than a satisfaction for him to have. Mr. Wells, however, is original in all things, and shows us in this story what a disadvantage it is to become invisible. He describes how, if a man becomes invisible, it does not follow that the clothes he wears become invisible also, and on this supposition has woven a story that will hold the reader with breathless interest from start to finish.

Figure 22. Original magazine advertisement for H. G. Wells's novel *The Invisible Man* (1897).

The story is now a familiar one: a scientist named Griffin discovers the secret to turning a living creature completely invisible, and he rashly tests his concoction on himself. However, his clothes do not turn invisible with him, and Griffin learns that invisibility can be as much a curse as a blessing because he is forced to bundle up in head-to-toe clothing while in public and to perform any clandestine missions fully naked. As his madness grows, Griffin attempts to convince a fellow scientist named Kemp to aid him in a megalomaniacal reign of terror against the country, and Kemp makes a desperate attempt to stop him. But how does Griffin achieve this invisibility? He explains the details to Kemp:

> If a sheet of glass is smashed, Kemp, and beaten into a powder, it becomes much more visible while it is in the air; it becomes at last an opaque white powder. This is because the powdering multiplies the surfaces of the glass at which refraction and reflection occur. In the sheet of glass there are only two surfaces; in the powder the light is reflected or refracted by each grain it passes through, and very little gets right through the powder. But if the white powdered glass is put into water, it forthwith vanishes. The powdered glass and water have much the same refractive index; that is, the light undergoes very little refraction or reflection in passing from one to the other.
>
> You make the glass invisible by putting it into a liquid of nearly the same refractive index; a transparent thing becomes invisible if it is put in any medium of almost the same refractive index. And if you will consider only a second, you will see also that the powder of glass might be made to vanish in air, if its refractive index could be made the same as that of air; for then there would be no refraction or reflection as the light passed from glass to air.[10]

Here we find very much the same sort of explanation from Wells that Fitz James O'Brien used in his story "What Was It?" Wells, through Griffin, notes that an object is largely visible due to the reflection and refraction of light from it. But optical science had progressed dramatically from the time of O'Brien: through the use of Maxwell's equations, it had been shown that there is always *some* reflected light whenever light illuminates an interface between media of two different refractive indices. When light goes from air into water, for example, there is always some light reflected back into air, because the index of refraction of air is approximately 1 and the index of water is 1.33.

Wells gets around this by achieving invisibility through "index matching." If two materials have the exact same refractive index, then there is no refraction, and consequently no reflection. As an example, Pyrex glass (often used for cookware) has a refractive index of 1.474, which is almost exactly the same as mineral oil that can be purchased in a drugstore. If one submerges a piece of Pyrex glass in a dish of mineral oil, the glass will seem to disappear into the liquid.

One difficulty with this explanation, even for a science fiction story, is that the refractive index of air is almost exactly the same as that of a vacuum. Any attempt to make an object invisible in air using index matching would have to match the refractive index of the object to the index of literal empty space. But here, Wells drew on the science of the time. In the novel, Griffin explains it thusly: "But the essential phase was to place the transparent object whose refractive index was to be lowered between two radiating centres of a sort of ethereal vibration, of which I will tell you more fully later. No, not those Röntgen vibrations—I don't know that these others of mine have been described. Yet they are obvious enough."[11] The public imagined that X-rays were somehow making objects and people invisible; Wells used this misconception and imagined a new set of rays that would literally turn an object invisible to the naked eye. This idea was implausible, even to the general public, but Wells achieved great success in science fiction by building his stories around a single implausible idea and making everything around it realistic. As he himself described it, "As soon as the magic trick has been done the whole business of the fantasy writer is to keep everything else human and real. Touches of prosaic detail are imperative and a rigorous adherence to the hypothesis. Any extra fantasy outside the cardinal assumption immediately gives a touch of irresponsible silliness to the invention."[12]

Clearly, the discovery of X-rays was a major inspiration for *The Invisible Man,* though there might have been other influences. In 1881, Edward Page Mitchell published his short story "The Crystal Man" in the *New York Sun.* The tale describes the troubles of a lab assistant named Flack who allowed himself to be used as a test subject for invisibility but became stuck in that condition. Unlike Wells's Griffin, however, Flack is furnished with a set of invisible clothes to preserve his dignity. Mitchell does not provide as many details of the invisibility process, other than to describe it vaguely as a chemical bleaching,

> "Now," continued Flack, "to the story of my undoing. The great histologist with whom it was my privilege to be associated, next turned his attention to another and still more interesting branch of the investigation. Hitherto he had sought merely to increase or to modify the pigments in the tissues. He now began a series of experiments as to the possibility of eliminating those pigments altogether from the system by absorption, exudation, and the use of the chlorides and other chemical agents acting on organic matter. He was only too successful!"[13]

Flack meets his tragic end from heartbreak: when he reveals his condition to the love of his life, she mocks and spurns him, leading him to take his own life.

Though the idea of bleaching a human to transparency may seem far-fetched, something close to this has been done in recent times. In 2001, Japanese researchers introduced a chemical reagent called Sca*l*e that can render biological samples transparent.[14] They were able to take sections of mouse brains and even unliving mouse embryos and render them see-through. In a statement that would make Edward Page Mitchell and H. G. Wells proud, they said, "We are currently investigating another, milder candidate reagent which would allow us to study live tissue in the same way, at somewhat lower levels of transparency."[15]

Wells would continue to write novels for the rest of his life, but his greatest contributions to science fiction came in that first decade between 1895 and 1905. After that, his work became more socially conscious and political, reflecting his efforts to make the world a better place and even bring about a utopian society. But his influence in science fiction persisted, and more invisible men would terrorize the public in literature as the decades passed.

One striking example was written by another famed author of scientific romances. Jules Verne, author of *Journey to the Center of the Earth* (1867) and *Twenty Thousand Leagues under the Sea* (1871), was inspired by Wells to write *The Secret of Wilhelm Storitz* in 1897. In the story, the titular Storitz uses the power of invisibility to seek revenge on the woman who spurned his advances and her family. The secret to invisibility is again intimated to involve Röntgen's mysterious rays. Verne died before the novel, his last, was published, and it appeared in 1910, after substantial revisions by his son. An English translation of Verne's original text finally appeared in 2011.[16]

An even more vicious invisible man appears in Philip Wylie's *The Murderer Invisible,* a novel from 1931. Wylie was another hugely influential author of science fiction; his novel *Gladiator* (1930), about a man who is imbued with super strength, is thought to be one of the inspirations for the comic book hero Superman, who first appeared in 1938. Wylie also wrote the classic novel *When Worlds Collide* (1933), an apocalyptic tale about humanity's struggle to survive when it is discovered that a rogue planet is approaching and will soon destroy the Earth.

The Murderer Invisible can be summarized by the question, "What would have happened if Griffin from *The Invisible Man* had been able to carry out his reign of terror?" In the novel, a mad scientist named Carpenter succeeds in giving himself personal invisibility and creates worldwide panic with a series of murders, bombings, and arsons. But Carpenter's plans are nearly undone before they begin when he finds that his bones do not vanish as rapidly as the rest of his body, leaving him a very visible living skeleton, just as a mob comes to his house to stop him: "Carpenter's rejoinder was frantic. 'You fool! I'm a man. A man like you. I've had an accident that makes me look this way. It isn't black magic. It's science.' The jaws wiggled. The skull turned from side to side. Some one hit him on the head with a club. He sank to the ground. Some one else kicked him. Men sat upon him, held him down with their feet."[17]

Of course, numerous movies have been made about invisible men, too. *The Invisible Man* (1933), directed by James Whale, closely follows the plot of Wells's novel. The film *Hollow Man* (2000), directed by Paul Verhoeven, is the story of a test subject who is made invisible but goes insane when the process cannot be reversed. Even more recently, *The Invisible Man* (2020), directed by Leigh Whannell, draws its inspiration from the same source but tells a very different story of obsession, envisioning its villain wearing a suit that produces an active form of invisibility. Wells's novel was a turning point in the history of invisibility physics, when the possibility of invisibility—and its dangers—entered the public consciousness, where it has remained to this day. After Wells, it was inevitable that scientists would ask the question: Is invisibility *really* possible?

9
What's in an Atom?

> At every step I found myself stopped by the imperfections of my instruments. Like all active microscopists, I gave my imagination full play. Indeed, it is a common complaint against many such, that they supply the defects of their instruments with the creations of their brains. I imagined depths beyond depths in nature which the limited power of my lenses prohibited me from exploring. I lay awake at night constructing imaginary microscopes of immeasurable power, with which I seemed to pierce through all the envelopes of matter down to its original atom.
>
> *Fitz James O'Brien, "The Diamond Lens" (1858)*

Though science fiction authors started considering the possibility of invisibility as far back as the 1850s, science did not begin to catch up until about sixty years later. The first steps toward scientific invisibility would come from a very unlikely source: attempts to answer the fundamental question, "What's in an atom?"

The idea of an atom predates the existence of formal science. The name "atom" comes from ancient Greece; in the fifth century BCE, the philosopher Democritus and his teacher Leucippus introduced the term with the concept that all matter is made from such fundamental indivisible components, surrounded by an emptiness called the "void." Similar ideas in fact appeared much earlier, in the eighth century BCE in India, starting with the Hindu sage Aruni. This idea of atoms was based entirely on philosophical arguments, without any experimental evidence to back them up. In ancient Greece, later philosophers largely rejected the atomist philosophy.

The idea of atoms finally began to enter scientific circles in the seventeenth century, and Isaac Newton speculated on a version of an atomic theory, without naming it as such, in later editions of his *Opticks:*

> Now the smallest Particles of Matter may cohere by the strongest Attractions, and compose bigger Particles of weaker Virtue; and many of these may cohere and compose bigger Particles whose Virtue is still weaker, and so on for divers Successions, until the Progression end in the biggest Particles on which the Operations in Chymistry, and the Colours of natural Bodies depend, and which by cohering compose Bodies of a sensible Magnitude.[1]

Atoms only entered the scientific mainstream, however, some one hundred years later. Ironically, at the same time that Thomas Young and others were recognizing that light acts more like a continuous wave than a particle, other researchers were finding evidence that matter consists of discrete particles rather than a continuous, infinitely divisible substance.

The major driver of this change of perspective was the English chemist John Dalton, who formally introduced an atomic theory in his volumes on *A New System of Chemical Philosophy,* which began publication in 1807. Dalton suggested that the chemical elements are made up of extremely small and indestructible particles that he called "atoms," and he argued that atoms of a single element, like oxygen, are all identical to each other. The origin of Dalton's atomic theory is curiously unclear, even to Dalton himself, who gave contradictory stories during his lifetime.[2] But he brought a major piece of experimental evidence to the discussion, which is now known as the law of multiple proportions. This law states that if two elements may be combined to form more than one compound, then the ratios of the masses of the second element will always be a whole number.

This is best illustrated with an example. Carbon can combine with oxygen to form two different chemical compounds, one of which requires more oxygen than the other. With 100 grams of carbon, the first compound could be formed with 133 grams of oxygen, while the second compound could be formed with 266 grams of oxygen. Dalton noted that the amount of oxygen in the second compound was exactly twice that in the first compound, suggesting that the first compound has one oxygen atom for every carbon atom, while the second has two oxygen atoms for every carbon atom. Today we recognize these compounds as carbon monoxide and carbon dioxide.

The law of multiple proportions was indirect evidence for the existence of atoms, with the rules of chemical combination strongly indicating that matter must come in discrete bits of stuff. But it did not directly show the effect of individual atoms, leaving open the possibility that the law could be explained by other hypotheses. Dalton's work led many researchers to believe in atoms, but the idea remained somewhat controversial and disputed throughout the nineteenth century.

One other experimental observation early in the century would turn out to be pivotal in the verification of atomic theory. In 1827, the Scottish botanist Robert Brown was studying small pollen grains in water when he noticed that minute particles emitted from the grains exhibited irregular, jittery motion in the water, almost as if they had a life of their own. Brown repeated the experiment with tiny particles of inorganic matter, confirming that the motion was not due to living organisms; however, he was unable to explain what caused it, and this "Brownian motion" would remain a puzzle for the better part of a century.

Researchers were slowly coming to accept the existence of atoms, but they were hard-pressed to explain in any detail the nature of an atom. The consensus view as the century progressed imagined atoms as rigid, impenetrable spheres, surrounded by a "field of force" that caused them to be attracted to some elements and repelled by others, thus creating the complexity of chemistry. But even this vague picture conflicted with experiments being performed. In 1844, none other than Michael Faraday published "A Speculation Touching Electric Conduction and the Nature of Matter," laying out some of the baffling observations that had been made relating to atoms.[3]

According to the conventional picture of atoms, a solid material consisted of a bunch of these rigid atom spheres packed closely together yet not touching, kept apart by the forces between them. But then, Faraday asked, why would some materials be strong conductors of electricity and others strong insulators? A conductor like gold or silver, he reasoned, must conduct electricity through its gaps, and thus space must be a conductor. However, looking at an insulator like shellac, one would argue the opposite: space must be an insulator.

The rigid sphere model, as described, naturally makes a reader imagine spheres that are packed closely together. But Faraday noted that, in chemical reactions, some materials *decrease* in volume as one adds more atoms to the mixture. If one adds oxygen and hydrogen to pure potassium, for example, to get hydrate of potassa, the hydrate has a smaller volume. If atoms were closely packed spheres, adding more atoms to a mixture would always increase the volume. Faraday concluded, somewhat presciently, that the hard spheres of atoms must be *much* smaller than the field of force that surrounds them. Faraday went so far as to endorse the views of the Ragusan atomist Roger Joseph Boscovich (1711–1787), who suggested that an atom is purely a field of force and has no hard sphere at all or, at most, a pointlike sphere at its center.[4]

Faraday was careful to label his discussion of atomic structure a "speculation," because there was in his time no known way to look directly at the structure of an atom to see what it was made of. Science fiction authors, like Fitz James O'Brien in his story "The Diamond Lens" (1858), fantasized about creating a microscope so powerful that it could peer down into the spaces between atoms and see a hidden universe of subatomic living creatures. But one consequence that followed from Thomas Young's wave theory of light was the recognition that a microscope cannot resolve any objects smaller than the wavelength of light itself. Visible light, with a wavelength of roughly one half of a millionth of a meter, is sufficient to peer at living cells and some of the structures within them but is woefully inadequate to pick out the details of atoms (which later would be found to be of a size roughly one tenth of one *billionth* of a meter).

As the nineteenth century progressed, tantalizing hints about the structure of the atom began to be unveiled. The first clue was the periodic table of elements, introduced by the Russian chemistry professor Dmitri Mendeleev in 1869. This table, a modern version of which is probably found on the wall of every science classroom in the world, was constructed by Mendeleev as he worked on a textbook of chemistry. As he tried to classify the various known elements according to their chemical properties, he realized that the elements, arranged in increasing atomic weight, form a periodic structure in their chemical properties. By leaving gaps in the table where no known element existed, he could predict the properties of these yet undiscovered elements. Many of these were discovered in short order after Mendeleev's publication, providing strong validation of his work. The periodic table indicated that there was some sort of underlying structure that related all the elements to each other, though what that structure was, nobody could yet say.

The next important hint was the discovery of the electron as the fundamental carrier of electric charge. X-rays had been discovered in 1895 by Wilhelm Röntgen through experiments with cathode rays carrying electricity; in 1897, the British physicist J. J. Thomson demonstrated experimentally that these rays were in fact a stream of negatively charged particles, each with a mass a thousand times smaller than the smallest atom. Thomson called them "corpuscles," but the name "electron" was later made standard. It was immediately recognized that these electrons were a fundamental building block of atoms, but again, nobody could yet say what role they played in atomic structure.

A third key hint was the discovery of radioactivity. In 1896, the French physicist Henri Becquerel was studying the phenomenon of phosphorescence, in which a material absorbs light and emits different radiation over a long time scale—it glows for an extended period of time. On learning of the discovery of X-rays, Becquerel wondered whether X-rays were emitted in phosphorescence, and he designed experiments to test this possibility. He wrapped photographic plates in heavy black paper—so no visible light could reach them—and placed a phosphorescent material on top of the plates. Leaving this arrangement in the sun, he postulated that the sunlight might excite phosphorescent X-rays, which could penetrate the paper and develop the photographic plates.

His hypothesis was seemingly confirmed when he used phosphorescent uranium salts, and he reported his results to the French Academy of Sciences in late February 1896. However, he then had a moment of pure serendipity: after preparing further plates for experiments, he found himself faced with a cloudy day and put the photographic plates, with the uranium salts on top of them, in a drawer for future experiments. Several days later, he decided to develop one of the photographic plates anyway, and to his surprise he found an even stronger image than he had found with his sunlight experiments. Further testing with nonphosphorescent uranium salts led him to a surprising conclusion: the uranium was emitting mysterious rays all on its own, a phenomenon we now refer to as radioactivity. Additional radioactive elements were quickly discovered, and it was found that three types of radiation were emitted, with very different properties; these were labeled "alpha" (α), "beta" (β), and "gamma" (γ) radiation, after the first three letters of the Greek alphabet.

Though it would take many years for the full significance of radioactivity to be understood, it was further evidence that there was something quite complicated going on within atoms and that an atom must itself consist of multiple components.

A final important hint to the structure of atoms was found in the emission and absorption of light by atoms. In 1814, the German optical physicist Joseph von Fraunhofer attached a prism to a telescope to separate out the colors of sunlight, as Newton had done over a century earlier. Looking at the spectrum of sunlight, Fraunhofer noted that the bright color spectrum was punctuated by thin dark lines, where apparently no light was emitted by the Sun at all (fig. 23).

These lines were later traced to individual atomic elements, and it was found that individual elements have certain special wavelengths at which they absorb and emit light. For instance, if one burns or electrically excites a particular element and uses a prism to separate out the colors, the result is a spectrum of isolated bright lines. Conversely, sunlight passed through a transparent sample of that element (in gaseous form, for instance) shows that the element absorbs light at those same spectral lines. The 574 dark lines that Fraunhofer saw in the spectrum of sunlight represent the absorption of the Sun's rays

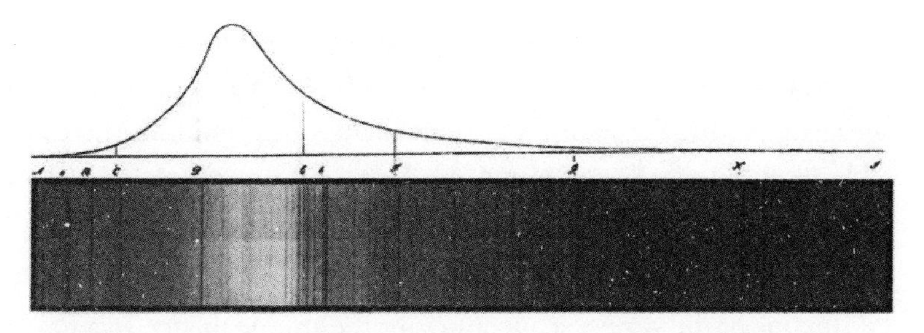

Figure 23. Joseph von Fraunhofer's original sketch of the solar spectrum, showing dark lines in the otherwise bright continuous spectrum (reproduced here in gray scale; be sure to look up the original version, in color, online).

by a variety of different elements in the Sun itself. Fraunhofer's observation gave rise to the field of spectroscopy—the determination of the chemical content of a material by its light emission—which is still a standard technique in science and engineering to this day.

These spectral lines of different elements clearly revealed something about the structure of atoms, but for the bulk of the nineteenth century nobody had any idea how to interpret them. In 1885, however, the Swiss mathematician Johann Jakob Balmer, at age sixty, looked specifically at the position of the spectral lines for the lightest atomic element, hydrogen, and found a very simple mathematical formula that allowed him to predict the positions of all of hydrogen's spectral lines for visible light. In 1888, the Swedish physicist Johannes Rydberg showed that Balmer's formula could be extended to also predict the spectral lines of hydrogen in the ultraviolet and infrared ranges and to estimate the wavelengths of the spectral lines of other atoms, too. This formula was the first quantitative hint at the structure of the atom.

By the beginning of the twentieth century, the scientific community was largely convinced of the existence of atoms and now had a handful of tantalizing but baffling clues as to their structure. However, they had no way to directly study the structure of atoms. In response, scientists did what they naturally do in such a situation: speculate wildly. Some of the greatest names of physics in the era, including multiple future Nobel Prize winners, would get in on the atomic speculation craze, which would last roughly a decade (fig. 24).

The earliest guess was made by the French physicist Jean Baptiste Perrin, one of the strongest advocates for the atomic theory. In a 1901 lecture on "the molecular hypothesis," he suggested that atoms are structured like the solar system, with a positively charged "sun" surrounded by planetlike electrons, "Each atom would consist, first, by one or more masses strongly charged with positive electricity, sort of suns whose positive charge is much higher than that of a corpuscle, and secondly, by a multitude of corpuscles, sort of small negative planets, the whole of masses gravitating under the action of electric forces, and the total negative charge exactly equal to the total positive charge, so that the atom is electrically neutral."[5] Perrin's hypothesis suffered from one significant weakness: the orbits of electrons around a positive sun would be unstable and prone to collapse under any significant perturbation. Our own planetary solar system appears stable because we (thankfully) do not have collisions with other solar systems; planetary atoms, colliding with each other regularly, would be knocked to pieces in short order.

J. J. Thomson, discoverer of the electron, addressed this limitation with his own model.[6] He imagined that the positive charge of the atom is smeared out as a positively charged fluid within which the electrons orbit. Thomson produced some thirty pages of calculations showing how this system could, under the right circumstances, provide stability for the orbiting electrons. This model became known as the "plum-pudding" model, as it could be explained as electron "plums" circulating in a positive "pudding." Thomson further demonstrated that electrons could be ejected from the pudding if their speed

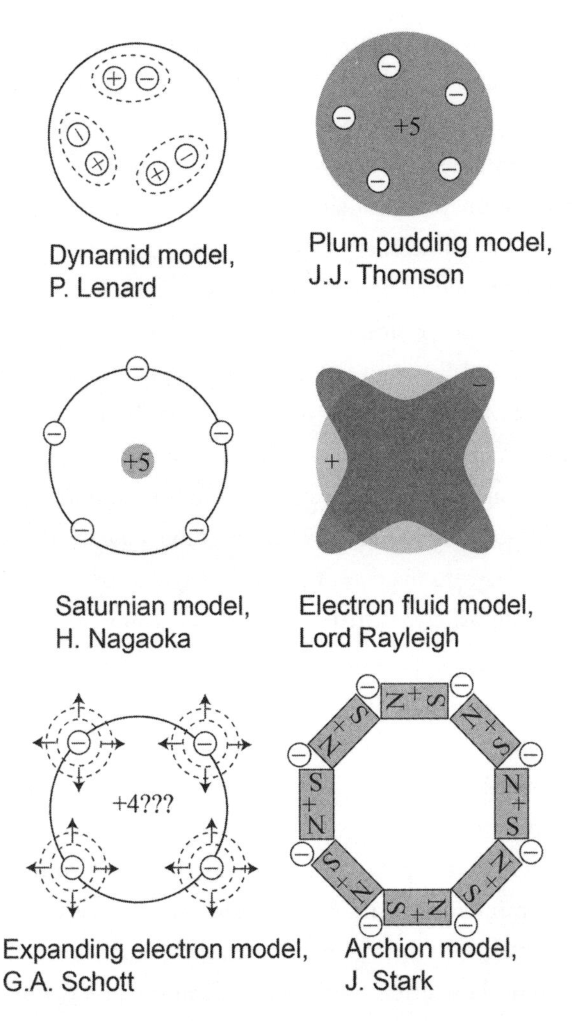

Figure 24. Various atomic models proposed in the early 1900s.

decreased below a critical threshold, producing a phenomenon that could be interpreted as radioactivity.

A different approach to the stability problem was suggested by the Japanese physicist Hantaro Nagaoka in 1904.[7] He kept the positively charged sun of Perrin's work, but suggested that the electrons in the atom are arranged in rings, like those around the planet Saturn. In 1859, James Clerk Maxwell had demonstrated theoretically that Saturn's rings are stable when slightly perturbed, oscillating rather than breaking apart. Nagaoka showed that the analogous oscillations of a Saturnian atom could produce something roughly similar to the spectral lines observed by Fraunhofer and others.

In 1906, another famous physicist leapt into the fray. Throughout his life, Lord Rayleigh made countless contributions to theoretical and experimental physics, focused primarily on fluid mechanics and optics, including an accurate explanation for why the sky is blue. Rayleigh would combine his strengths in a modification of Thomson's plum pudding model.[8] Thomson's model assumed a very small number of electrons in an atom, and Rayleigh asked what would happen if the opposite were true: What if the atom contains so many electrons that the collection can be considered a fluid? In his calculations, Rayleigh found that this sea of electrons could itself vibrate, like a bowl of Jell-O, and that these vibrations could result in discrete spectral lines. However, the lines Rayleigh calculated did not match the Rydberg formula at all.

Also in 1906, the physicist James Jeans, a British cosmologist and author of numerous technical and popular science books, offered another twist on Thomson's model. Astutely, Jeans noted that, based on the physical quantities involved in Thomson's model, it is mathematically impossible to derive a quantity that depends on the wavelength or, equivalently, the frequency.[9] Thus Thomson's model by itself simply cannot produce the discrete spectral lines of the elements. In this case, Jeans suggested that electrons might not be pointlike particles, as previously assumed, but elastic spheres of finite size, and that these electron spheres might vibrate, producing line spectra.

One other researcher indulged in atomic speculation in 1906. The

British mathematician George Adolphus Schott agreed with Jeans that something was missing from Thomson's model, and he also concurred that the answer might be an electron of finite size. But Schott argued that the vibrational frequencies of the electron arise because the electron is constantly trying to grow in size, and its growth is resisted by the hypothetical aether through which electromagnetic waves travel.[10] In his calculations, Schott further noted that his model predicts an attraction between electrons, and he postulated that this could be the force of gravity. Schott's theory was not seriously considered by other researchers, but it would not be the last time that Schott would produce a truly imaginative solution to a problem, as we will see.

The explosive interest in atomic models in 1906 was not coincidental. In 1905, a then relatively unknown physicist named Albert Einstein published a paper with the translated title "On the Motion of Small Particles Suspended in a Stationary Liquid, as Required by the Molecular Kinetic Theory of Heat."[11] In this paper Einstein argued that the irregular Brownian motion of small particles in a liquid could be explained by the intermittent collision of the visible particles with the unseen atoms of the liquid. This idea had been suggested previously, but Einstein presented a detailed mathematical analysis of the hypothesis, making predictions that could be tested experimentally, and his work renewed interest in understanding atomic structure. By 1910, Jean Perrin had verified Einstein's predictions concerning Brownian motion, and the last of the atom disbelievers surrendered in the face of overwhelming evidence.

None of the researchers making speculations knew which observations about atoms would be most important in unlocking their secrets. The aforementioned researchers centered their attention on explaining the Balmer and Rydberg formulas, but others focused on understanding the structure of the periodic table.

The Hungarian-German physicist Philipp Lenard did extensive work on cathode rays in the late nineteenth century for which he would later win the Nobel Prize in physics, and he had noted that the ability of a material to absorb electrons depends on the mass of the material and hardly at all on its specific chemical properties. This suggested to him that atoms are all made of the same fundamental pieces and that the only difference between different elements is the presence of more or less of these fundamental pieces. In 1903 he postulated that the fundamental building block of all elements is what he called the "dynamid": a positive and negative charge bound together.[12] In his model, the atomic mass of an atom is simply proportional to the number of dynamids present—that is, a hydrogen atom is a single dynamid, helium would evidently be four dynamids, and so on. The binding force that holds these dynamids together was not explained. This model could roughly explain the structure of the periodic table but not the line spectra of atoms.

The most imaginative model was introduced in 1910 by German physicist and future Nobel Prize winner Johannes Stark.[13] Stark, also seeking to explain the structure of the periodic table, proposed a fundamental unit of positive charge called the archion, which he imagined as essentially a tiny bar magnet possessing a positive electric charge. These positive charges, which would normally be repelled strongly from each other, are held together by the magnetic force and the negatively charged electrons. Stark imagined that these archions formed closed rings of bar magnets, with adjacent North and South poles lined up. Different atomic elements correspond to rings of different size. Stark's model hinted at the meaning of the periodic table but was limited in that it could not explain the Rydberg formula.

It is important to note that these attempts to describe the atomic structure were not idle speculation but rather an international scientific brainstorming session, acted out in writing through multiple scientific journals in multiple countries. Each scientist shared their incomplete idea with the hope that others could build on it and come closer to the truth.

But every model proposed at that time had a fundamental flaw. Maxwell's equations predict that any accelerating electrical charge—such as one moving in a circular path—will emit radiation. Today, this has practical applications: every radio broadcast tower and cellphone uses oscillating electrical currents to generate radio waves. At Argonne National Laboratory in Illinois, the Advanced Photon Source sends electrons around a circular storage ring with a 1,100-meter circumference at nearly the speed of light, and the accelerating electrons produce X-rays that can be used for a variety of research projects.

Every aforementioned atomic model explicitly or implicitly required electrons to be moving in a circular orbit around the center of the atom. Researchers quickly calculated how much radiation these atomic electrons must emit and found that the electrons should expend all their energy and collapse into the core in a fraction of a second. For any atomic model to survive, it would have to explain why stable atoms exist at all.

For the first decade of the twentieth century, there was no significant attempt to solve this conundrum. In 1910, however, the Austrian-Dutch theoretical physicist Paul Ehrenfest published a short paper with the translated title "Irregular Electrical Movements without Magnetic and Radiation Fields."[14] In it, Ehrenfest argued that it is possible, under the right circumstances, for an *extended* distribution of electrical charge, such as the vibrating electron sphere of James Jeans, to accelerate without producing any radiation. This insight would later be recognized as one of the first major scientific papers related to the physics of invisibility.

At the time of this work, Ehrenfest was still relatively new to science. He had developed an interest in statistical mechanics, which applies mathematics to describing the behavior of large groups of particles, such as molecules in liquids and gases. Statistical mechanics underwent a surge of interest in part due to Einstein's explanation of Brownian motion, which showed the power of using statistics to quantify physical problems. Ehrenfest got his PhD in 1904, and by 1906 he was working on a review article about statistical mechanics with his wife, Tatyana, also a talented mathematician. Ehrenfest would eventually be recognized as one of the foundational researchers in the field. With this work, thoughts on atoms and their behavior would never be far from Ehrenfest's mind, and he was clearly aware of the issues with the existing atomic models.

Ehrenfest began his paper with a pair of examples using one of the most powerful tools in the toolkit of a theoretical physicist: symmetry. First, he imagined an infinitely large, flat sheet of material with electric charge uniformly distributed across its surface. While the sheet is not in motion, its electric field must point perpendicular to the sheet, because the sheet looks the same at every point along its surface: it is symmetric along the surface. He then imagines the sheet oscillating up and down (fig. 25). This sheet of electric charge is accelerating, and one would therefore expect it to produce radiation. However, any radiation produced must also travel perpendicular to the sheet, and any magnetic field created must travel perpendicular to the sheet. This means that the only way that an electromagnetic wave can be generated by this system is with the electric field and the magnetic field both pointing along the direction the wave is traveling. But, as Maxwell showed, electromagnetic waves are transverse and always have the electric and magnetic waves perpendicular to the direction of wave propagation. Therefore, Ehrenfest argued, no radiation is generated by this system. The electric and magnetic fields produced by the oscillating sheet are localized to the immediate area of the sheet.

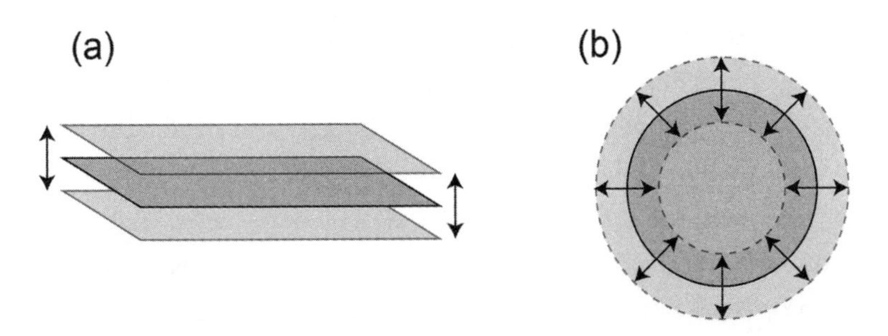

Figure 25. Paul Ehrenfest's simple models: (a) an infinite plane vibrating up and down, and (b) a pulsating sphere.

Ehrenfest himself admitted that this example is somewhat artificial, because it requires a sheet of electrical charge of infinite length and width, something that does not exist in nature. So he presented a second example: a spherical shell uniformly coated with electrical charge, with its radius increasing and decreasing. One can imagine this example as a spherical balloon, dusted with electrical charge, being rapidly inflated and deflated. All the motion of the charges is in the radial direction, toward or away from the center of the sphere; again, by symmetry, the only direction that the electric and magnetic fields can point is in the radial direction. Furthermore, the only direction the wave can propagate is in the radial direction. Again, Ehrenfest stated, we have a situation where the only possible direction for the electric field, magnetic field, and wave propagation is the radial direction; but since a wave can propagate only if all three of these quantities are perpendicular, no wave can propagate. The pulsating sphere is also radiationless.

Symmetry arguments in physics can be difficult to grasp, so I'll present an analogous argument from philosophy, known as Buridan's ass, named after the fourteenth-century French philosopher Jean Buridan. Imagine a donkey placed equidistant between two identical piles of hay. Buridan and others have argued that the donkey will go to whichever pile of hay is closer. Because both piles are equally close, the donkey cannot decide and will stay in place and starve. This paradox is used in philosophy in discussions of the concept of free will.

We may use a reasoning similar to Buridan's ass in explaining Ehrenfest's symmetry arguments. In order for an electromagnetic wave to propagate away from the vibrating plane, it must have transverse electric and magnetic fields—that is, those fields must point somewhere along the surface of the plane. But there is nothing in the setup of the system that "chooses" a direction for those electric and magnetic fields to point: all motion is perpendicular to the plane. Without some part of the system choosing a direction, no electromagnetic wave can be generated.

Ehrenfest did not rely solely on these two simple examples. He also showed how to use Maxwell's equations to construct—theoretically, at least—a wide variety of extended distributions of accelerating electrical charges that would not produce radiation. Far from being an isolated quirk, the physics of radiationless oscillations could be seen as a fundamental part of electromagnetic wave theory. Though pointlike electric charges will always produce radiation when accelerating, extended volumes of charge have the possibility of accelerating without producing radiation.

Ehrenfest's theory did not get much attention, and it is still largely unknown in the broader physics community. This seems to be the result of bad timing—at the same time Ehrenfest was working on his results, new experimental data about the structure of the atom would change the conversation and usher in an entirely new era of physics.

Back in 1895, Philipp Lenard had demonstrated that electrons could be transmitted through a thin quartz window. From this, he concluded that there must be significant space between the atoms of the window in order for the electrons to pass through. There was a greater significance to this work: it showed that electrons, and other small particles, had the potential to probe the structure of matter, something impossible to do with light.

New Zealand physicist Ernest Rutherford decided to explore this possibility. In 1895, Rutherford was awarded a fellowship that allowed him to travel and do postgraduate research at the Cavendish Laboratory at the University of Cambridge, putting him in a research position at the time when X-rays, electrons, and radioactivity were all discovered. He worked under J. J. Thomson and was therefore present for the discovery of the electron. In 1898, Rutherford accepted a professorship at McGill University in Canada, and there he did extensive research into radioactivity and X-rays. One of his notable discoveries was the observation that there are different forms of radioactivity, and Rutherford is the person who coined the terms "alpha" (α) and "beta" (β) radiation. He made many other significant discoveries related to radioactivity and in 1907 moved back to England to accept a position at Victoria University of Manchester. A year later, he was awarded the Nobel Prize in chemistry "for his investigations into the disintegration of the elements, and the chemistry of radioactive substances."[15]

At Manchester, Rutherford decided to study the interactions of alpha particles with matter. Alpha particles had already been identified as helium atoms stripped of their electrons: basically tiny but massive balls of positive electric charge. An alpha particle passing through a thin sheet of matter would be expected to be slightly deflected, due to interactions with the atoms of the material. Under Thomson's plum pudding atomic model, it was expected that this deflection would be very small—the diffuse "pudding" would provide little resistance to the high-speed alphas, and the electrons in the pudding would be of too low a mass to be a significant barrier. Rutherford and his assistant, Hans Geiger, used a radioactive source to fire alpha particles at thin sheets of gold foil. Gold was used because it could be hammered very thin and it could be hammered to a very precise thickness. Rutherford and Geiger were probing the internal structure of atoms by firing alpha particles through them.

For the initial experiments, Rutherford and Geiger looked for alpha particles deflected directly behind the gold foil. Their results showed that the alphas were mostly scattered by an angle of one degree, seemingly confirming Thomson's model. In 1909, Rutherford suggested a new experiment; as he later described it,

> One day Geiger came to me and said, "Don't you think that young Marsden, whom I am training in radioactive methods, ought to begin a small research?" Now I had thought that too, so I said, "Why not let him see if any α-particles can be scattered through a large angle?" I may tell you in confidence that I did not believe that they would be, since we knew that the α-particle was a very fast massive particle, with a great deal of energy, and you could show that if the scattering was due to the accumulated effect of a number of small scatterings the chance of an α-particle's being scattered backwards was very small. Then I remember two or three days later Geiger coming to me in great excitement and saying, "We have been able to get some of the α-particles coming backwards. . . . " It was quite the most incredible event that has ever happened to me in my life. It was almost as incredible as if you fired a 15-inch shell at a piece of tissue paper and it came back and hit you.[16]

According to Thomson's plum pudding model of the atom, no alpha particles should have been deflected backward. Rutherford pondered the experimental data, made some calculations, and concluded that the backward-going alphas were possible only if the atom consisted of a very small, very heavy positive nucleus, around which the electrons presumably orbited. Perrin's original idea of a planetary atom, with a positive sunlike nucleus surrounded by planetlike electrons, had seemingly been vindicated. Perrin graciously and accurately gave full credit to Rutherford, however. In Perrin's own Nobel Prize acceptance speech in 1926, he said:

> I was, I believe, the first to assume that the atom had a structure reminding to that of the solar system where the "planetary" electrons circulate around a positive "Sun," the attraction by the centre being counterbalanced by the force of inertia (1901). But I never tried or even saw any means of verifying this conception. Rutherford (who had doubtless arrived at it independently, but who also had the delicacy to refer to the short phrase dropped during a lecture in which I had stated it) understood that the essential difference between his conception and that of J. J. Thomson was that there existed near the positive and quasi-punctual Sun, enormous electrical fields as compared with those which would exist inside or outside a homogeneous positive sphere having the same charge, but embracing the whole atom.[17]

With Rutherford's discovery, and its subsequent publication in 1911, a new era of atomic physics had begun.[18] There was no longer any question about the general structure of the atom: an atom consisted of a tiny, heavy, positively charged nucleus surrounded by orbiting electrons. Soon it would be determined that the nucleus itself consists of multiple densely packed particles; radioactivity could be explained as the breakdown of heavier nuclei. But Rutherford's atom, with electrons rapidly orbiting the nucleus, brought renewed attention to the question that Ehrenfest had attempted to answer: Why doesn't an atom radiate?

This question would take several years to resolve, and the answer would lie in something that none of the earlier atomic modelers had considered: an entirely new system of physics, which we now refer to as the quantum theory. Quantum physics would render Ehrenfest's solution unnecessary to explain the structure of the atom, but Ehrenfest's solution would be revived years later to attempt to disprove that very same quantum physics.

10
The Last of the Great Quantum Skeptics

I have a modification of that first crude shield completely surrounding my body. It's really a flexible, ventilated metal mesh, not very heavy, with interstices so fine that the human eye cannot see them. I can breathe comfortably, perspire normally and move freely. But you can't see either me or my suit of mesh. The mesh carries a certain fine electric current, from special batteries at my waist, which kicks the light photons along as they arrive. For instance, a light photon striking my back is kicked right through my body to my front, and there radiated—as though I hadn't been in its way in the first place.

And that's the reason I'm invisible, because light goes through me, even more perfectly than light penetrates glass.

Eando Binder, "The Invisible Robinhood" (1939)

The seeds of a new type of physics had been planted years before Rutherford discovered the atomic nucleus. Starting in the mid-1800s, researchers had been studying the light emission of bodies that are hot enough to glow, such as the Sun, stars, and electric stove elements. It was quickly recognized that all bodies of this form possess a similar overall emission spectrum; an example of this is shown in the upper curve in Joseph von Fraunhofer's sketch of the solar spectrum, showing the brightness of the Sun as a function of wavelength. This emission spectrum, neglecting the dark spectral lines due to the absorption of specific elements, appeared to have a universal shape that depends only on the temperature of the object. The peak of the shape moves to shorter wavelengths with increasing temperature. Thus, a "red-hot" object is a lower temperature than a "white-hot" object.

To understand this thermal emission spectrum, physicists imagined an idealized hypothetical object that perfectly absorbs any radiation illuminating it—a "blackbody"—and then attempted to calculate how this object would reemit electromagnetic radiation. Many researchers attempted to derive the emission spectrum of blackbodies from fundamental physical principles, but for most of the nineteenth century no solution could be found that exactly matched the experimental data.

Near the end of the century, the solution was found at last by the German theoretical physicist Max Planck (1858–1947), then a professor in Berlin. Planck made progress by attacking the problem in the opposite direction from that of his contemporaries. Where other researchers had started from fundamental physical principles and tried to derive the mathematical form of the blackbody spectrum from those principles, Planck first hunted for a mathematical form for the spectrum, in agreement with the experimental data, and then sought a physical theory that would produce the mathematical result.

Planck started his work on blackbody radiation in 1894. By 1900, he had developed a formula for the blackbody spectrum that not only matched the experimental data but matched it perfectly. He had struck on the correct answer to the problem but did not have any idea how to derive that answer.

Planck's initial attempts to explain his results using the known physics of the era met with failure. A blackbody may be viewed as the combination of two systems interacting with each other: a collection of "oscillators," presumably atoms, vibrating and giving off electromagnetic radiation, and the electromagnetic radiation itself, which can be reabsorbed by the atoms. Using Maxwell's wave theory of electromagnetism, Planck predicted that all the energy of the atoms should inevitably convert into electromagnetic waves, leaving the matter cold and depleted. This was completely at odds with what was seen in experiments. In hindsight, it is telling that Planck's trouble was very similar to the difficulty that physicists of the time had explaining why an atom does not immediately radiate away all of its energy as electromagnetic waves; Planck would unknowingly take the first step in solving both problems.

Planck finally found a solution by making an unprecedented assumption about the vibrating atoms of the blackbody: he assumed that they could only emit electromagnetic energy in discrete amounts, which he called quanta, and that the fundamental unit of one of these quanta was directly proportional to the frequency of the electromagnetic wave. As Planck himself described it years later,

> Briefly summarized, what I did can be described as simply an act of desperation. By nature I am peacefully inclined and reject all doubtful adventures. But by then I had been wrestling unsuccessfully for six years with the problem of the equilibrium between radiation and matter and I knew that the problem was of fundamental importance to physics; I also knew the formula that expresses the energy distribution in the normal spectrum. A theoretical interpretation therefore had to be found at any cost, no matter how high.[1]

Planck had basically introduced new physics, including a new fundamental constant of nature that would be labeled in equations as h and referred to as Planck's constant. A quantum of energy for a given frequency would then be given by Planck's constant times the frequency. With energy labeled E and frequency labeled with the Greek letter ν, the equation for a quantum of light energy is then $E = h\nu$.[2]

Planck himself did not view his breakthrough as the dawn of a new era of physics, however; of his new quanta of energy, Planck commented, "This was purely a formal assumption and I really did not give it much thought except that no matter what the cost, I must bring about a positive result."[3]

The solution to another nineteenth-century physics puzzle would dramatically change this view of Planck's work. The puzzle is known today as the photoelectric effect and was discovered in 1887 by Heinrich Hertz as he did experiments to verify Maxwell's prediction of electromagnetic waves. To detect the electromagnetic waves, Hertz used a wire antenna with a gap in its center, called a spark gap; the incoming waves would induce an electrical current in the wire that would cause an electrical spark to leap across the gap. To see the spark better, Hertz placed his antenna in a darkened box; to his surprise, the sparks became weaker. Investigations showed that exposure to ultraviolet light was responsible for making stronger sparks: it liberated electrons from the surface of the metal that it was illuminating.

This result was perhaps not surprising in itself; it was well known by that point that electromagnetic waves carry energy and momentum, and it was reasonable to expect that these waves could provide enough of a kick to free electrons from a metal surface. Further experiments by Philipp Lenard in 1902, however, showed that the photoelectric effect had properties not easily explained by the wave theory of light.[4] Lenard measured the speed of electrons ejected from the surface of the metal and found that the speed was independent of the intensity of the ultraviolet radiation used. Furthermore, he found that the speed of electrons does depend on the frequency of the light used: higher frequencies correspond to faster electrons. According to the wave theory of light, it was expected that the electron speed would depend on the intensity of light because that would correspond to more energy, which could then be imparted to the electrons. It was also expected that the electron speed would not depend at all on the frequency of light. Lenard and other physicists struggled to find a wave theory that could account for the curious properties of the photoelectric effect.

The correct explanation was found and published by Albert Einstein in 1905, and it would revolutionize our understanding of light and matter. His paper, with the English translated title "On a Heuristic Viewpoint Concerning the Production and Transformation of Light," was the first of three major papers that he published in 1905.[5] The second paper, as we have seen, was an explanation of Brownian motion. The third paper would introduce his theory of special relativity, which we will discuss later.

In his explanation of the photoelectric effect, Einstein revised one hundred years of thinking about the nature of light. Whereas Thomas Young had demonstrated that light acts like a wave, Einstein argued that the photoelectric effect could most readily be explained if light also acts like a particle. In particular, he suggested that the light quanta that Planck introduced are not just a mathematical convenience but a fundamental property of light. The individual particles of light would eventually be called photons. In the photoelectric effect, individual electrons are therefore knocked off of a metal surface by individual photons. One can imagine the electrons as balls on a billiard table, and a photon as the cue ball, which strikes the billiard balls and sets them into motion. Because the energy of the photon depends on its wave frequency, a higher frequency photon ejects an electron at a higher speed. Because only a single photon interacts with each electron, the intensity of light (number of photons) affects only the number of electrons ejected, not their speed.

Einstein was not arguing that light only acts like a particle, as Newton had two hundred years earlier: he instead argued that light has both wave *and* particle properties, sometimes acting like a wave and sometimes acting like a particle. When light is propagating through space or traveling through a medium, it travels as a wave; when it is absorbed at a detector, it is absorbed like a localized particle. This idea of wave-particle duality is a foundational concept of quantum physics, and physicists still argue to this day about what, exactly, it means.[6]

The physics of the photoelectric effect would curiously be used in a unique science fiction explanation of invisibility. In the story "The Invisible Robinhood" (1939), by Eando Binder (the pen name of Earl AND Otto Binder), an early superhero creates a metal suit that uses electricity to kick photons through it from front to back, rendering the wearer invisible. The photoelectric effect is not mentioned explicitly, but the influence is unmistakable.

Einstein's hypothesis of light quanta was not readily accepted. Though the mathematical equations that Einstein produced ended up agreeing well with experiments, his use of new physics was viewed as too radical by many of the preeminent physicists of the time. The list of skeptics included Max Planck, who had first introduced quanta of light in the first place. The consensus thinking is captured well by physicist Robert Millikan, who in 1916 said,

> It was in 1905 that Einstein made the first coupling of photo effects and with any form of quantum theory by bringing forward the bold, not to say the reckless, hypothesis of an electromagnetic light corpuscle of energy $h\nu$, which was transferred upon absorption to an electron. This hypothesis may well be called reckless first because an electromagnetic disturbance which remains localized in space seems a violation of the very conception of an electromagnetic disturbance, and second because it flies in the face of the thoroughly established facts of interference.[7]

In short: the wave theory of light had been thoroughly proven over the past hundred years, and the notion of light as a localized particle seemed completely contradictory to the idea of light as a nonlocalized wave. Most physicists saw the introduction of new physics as creating more problems than solutions.

One scientist, however, was willing to make the quantum leap, so to speak, and go even further than Einstein: the Danish physicist Niels Bohr. Born in Copenhagen in 1885, Bohr enrolled as a physics undergraduate at the University of Copenhagen in 1903, where his father was a professor of physiology. He quickly distinguished himself, winning a gold medal competition presented by the Royal Danish Academy of Sciences and Letters in 1905. The prize problem was to investigate a method for measuring the surface tension of liquids; Bohr did the experimental work in his father's laboratory because the physics department had no lab of its own.

Bohr continued at Copenhagen to earn his master's degree and, in 1911, his PhD. His thesis was on the behavior of electrons in metals, and he traveled to England on a fellowship in that same year to meet with important scientists, including J. J. Thomson, the discoverer of the electron. He evidently failed to impress Thomson, but he also met Ernest Rutherford and was invited to spend time working with Rutherford in Manchester at just the time that Rutherford had discovered the existence of the atomic nucleus. Bohr therefore found himself at the center of activity in investigations of the structure of the atom.

He returned to Denmark in 1912 with atoms very much on his mind. He became a lecturer at the University of Copenhagen that same year and earned the equivalent of an associate professor rank in 1913. By mid-1913, he had come up with a new model for atomic structure, which he published in three papers (now known as "the trilogy") in July, September, and November that year.

Previous researchers, faced with the choice between keeping Rutherford's nuclear atom or keeping classical electromagnetism, opted to keep electromagnetism and look for other atomic models. After all, classical physics had predicted that a planetary atom like Rutherford's must radiate away all its energy in a very short period of time. In his first paper, Bohr made the opposite choice: he argued that Rutherford's picture is correct and that the laws of electromagnetism must be different for objects as small as an atom.[8]

Bohr then made two key assumptions: (1) light is emitted and absorbed by electrons orbiting the atom in the discrete corpuscles (photons) predicted by Einstein; and (2) an electron can only orbit at certain special distances from the nucleus. Bohr called these states of the electron stationary states, meaning that they are stable positions for electrons to orbit. An atom in one of these states does not emit light radiation as Maxwell's theory predicts.

Bohr argued that electrons can only "leap" from one special orbit to another; if they drop from an outer orbit to an inner orbit, they will release a photon in the process. The energy of the photon, and hence its frequency, is determined by how much energy the electron gives up in making the transition. A photon of an appropriate frequency can also be absorbed by an atom, causing the electron to leap from an inner orbit to an outer orbit (fig. 26).

What is the origin of these special orbits? Bohr argued that the angular momentum of the electron in orbit around the nucleus, its momentum of rotation, can only take on discrete values, just as light of a given frequency can only have energy in discrete multiples of $E = h\nu$. Bohr was, in essence, arguing not only that light energy is quantized in discrete amounts but that the angular momentum of electrons is quantized. A single quanta of angular momentum is given by Planck's constant h, indicating that both the Bohr atom and Einstein's photons are related to the same new fundamental physics.

As one can see, Bohr's model is filled with many assumptions, and it would have been easy for researchers to dismiss it, except: it matched

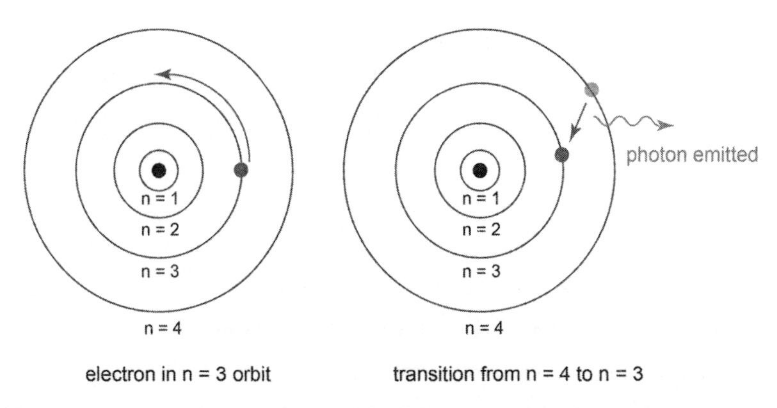

Figure 26. Niels Bohr's model of the atom. Electrons orbit at fixed distances, labeled by a number "n." When they transition from a higher orbit to a lower orbit, they emit a photon.

the Balmer and Rydberg formulas for light emission by a hydrogen atom almost perfectly. Bohr's discovery broke down a lot of resistance to the idea that new physics must be used to explain the atom; he would win the 1922 Nobel Prize in physics for his accomplishment.

One major conceptual question remained to be answered: Why is the angular momentum of electrons quantized, or more generally, why do electrons possess stationary states at all? The explanation would be presented by the French physicist Louis de Broglie in 1924, as his PhD thesis. De Broglie, born in 1892 into an aristocratic family, originally intended to pursue a career in the humanities. He was drawn into physics, however, through interactions with his brother Maurice, a physicist performing investigations of X-rays. De Broglie earned a physics degree in 1913 and then in 1914 was drawn into service in World War I, where he worked on developing radio communications for the military. On being released from duty in 1919, he immediately returned to those physics problems that the war had prevented him from studying, working toward a PhD at the University of Paris.

In his early work with Maurice, Louis de Broglie had studied both the photoelectric effect and the properties of X-rays. Through the photoelectric effect, he was well aware of the existence of photons and the wave-particle duality of light. As many others had observed before him, de Broglie further noted that X-rays, an electromagnetic wave, act a lot like particles due to their very short wavelength. This observation set de Broglie down a revolutionary path: just as light may be considered a wave with particlelike properties, he reasoned that matter might consist of particles that possess wavelike properties. If the wavelength of an electron is very small, it would behave under most circumstances like a particle. As he would later reflect in 1929 in his Nobel Lecture, "On the other hand the determination of the stable motions of the electrons in the atom involves whole numbers, and so far the only phenomena in which whole numbers were involved in physics were those of interference and of eigenvibrations. That suggested the idea to me that electrons themselves could not be represented as simple corpuscles either, but that a periodicity had also to be assigned to them too."[9] De Broglie envisioned the electron as a vibration spread out around the circumference of the atom. Young had observed that in an organ pipe, only certain waves with specific wavelengths can "fit" into the pipe; similarly, de Broglie reasoned that the electron wave can only fit in certain circular paths around the nucleus. With this wave picture, de Broglie was able to derive the angular momentum condition of electron orbits that Bohr had merely assumed (fig. 27). He successfully defended his PhD thesis in 1924, and by 1927, interference experiments had demonstrated the wave properties of the electron. The era of quantum physics was fully underway: it was now appreciated that everything in nature, both light and matter, possesses wave-particle duality.

The new quantum theory finally answered the question that had long plagued physicists: Why doesn't the electron radiate as it orbits

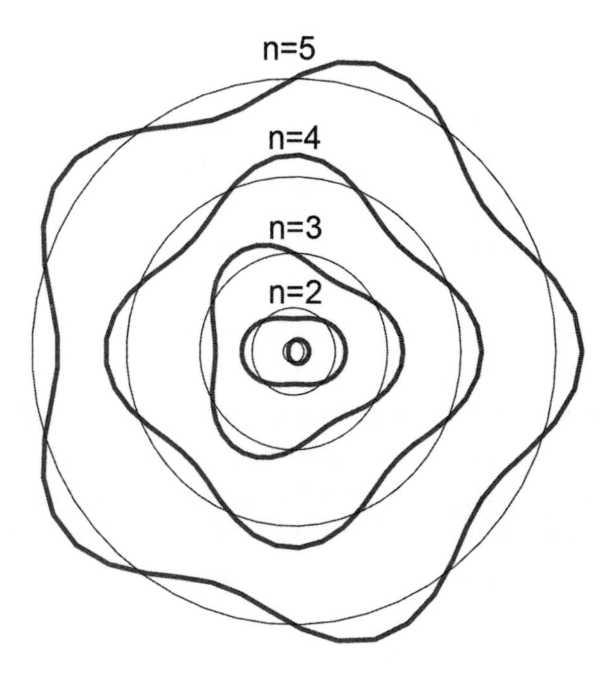

Figure 27. Louis de Broglie's electron waves around the circumference of the atom of different orders, which are labeled by a number "n."

around the nucleus? The answer is that an electron in one of its stationary states does not orbit at all; it is effectively a spread-out cloud enclosing the nucleus that does not move—a standing wave, like the sound waves in an organ pipe. Ehrenfest's hypothesis of radiationless oscillations of electric charges turned out to be unnecessary in the quantum theory.

Quantum physics, like all new theories, had many skeptics during its inception. The early work relied on a number of assumptions—quantized light, stationary orbits, and so forth—that appeared to many to cause more problems for physics than they solved.[10] As research continued and became more precise, most of these skeptics changed their views in the face of the increasing volume of experimental evidence and theoretical arguments that verified the quantum theory. Though questions remained even after de Broglie's work—and some remain to this day—most physicists came to embrace the new and revolutionary view of light and matter.[11]

At least one theoretician would remain unconvinced of the reality of quantum physics. The British physicist and mathematician George Adolphus Schott (1868–1937) would spend much of his career trying to prove that Maxwell's classical theory of electromagnetism can be used to derive all of the observed properties of atoms. Though he was unsuccessful in this endeavor, he developed new theoretical results relating to Ehrenfest's radiationless motions, bringing physics closer to thinking about the possibility of invisibility.

Relatively little is recorded of Schott's early life. He was born in Bradford, England, and did his early education there before proceeding to Trinity College, Cambridge, in 1886 to pursue his degree. He earned his bachelor of arts degree in 1890 and in 1893 became a lecturer in physics at the University College of Wales, Aberystwyth. He showed an early interest in the theory of electromagnetism, and his first published scientific paper, "On the Reflection and Refraction of Light," appeared in 1894.[12]

As we have seen, Schott joined in the craze of atomic speculation that infused the scientific community in 1903. His paper of 1906 proposing an expanding electron was his second published work. He continued studies of his atomic model as well as other atomic models for the rest of the decade, focusing much of his attention on how an atom can be stable without emitting radiation. In a paper from 1907, while discussing how atoms may consist of different rings of electrons orbiting at different distances from their center, he made the following telling statement: "Any two groups, on account of their permanent motion, disturb each other and emit waves, whose energy is a drain on the energy of the groups; hence the system is not permanent. But the drain of energy, and consequent change in structure, can proceed extremely slowly, provided all the most powerful waves due to perturbations are destroyed by interference between the several electrons of each group."[13] In his famous two-slit experiment, Thomas Young demonstrated that the light from the two slits can interfere and that at some points their waves can completely cancel out in the phenomenon of destructive interference. Analogously, Schott suggested that the waves emitted from two orbiting rings of electrons could largely if not completely cancel out, causing the atom to emit little to no radiation. He was suggesting a type of interference never seen before, and he had no mathematics to back up his intuition, but he was progressing toward a truly remarkable discovery, as we will see.

Schott made a scientific name for himself with his detailed mathematical studies of electron radiation, quickly distinguishing himself as a leading expert on the subject. Based on his interest and strength in the mathematical side of the problem, he became a lecturer in mathematics at Aberystwyth in 1909 and was promoted to the chair of the department in 1910.

After all of Schott's efforts to understand atomic structure using electromagnetic theory, Bohr's model of 1913 likely came as a shock. Schott found the introduction of an entirely new theory of matter to be troubling, and he spent the rest of his life trying to show that the curious properties of the atom, especially the radiationless orbits of Bohr, could be explained entirely using Maxwell's theory. As he would say of Maxwell's work in a paper published in 1918, "Doubtless no one will be willing to renounce so useful a theory as this is until much stronger reasons are forthcoming." It should be noted that he did not deny the validity of Bohr's work, as he considered the results "so extraordinarily exact that a considerable substratum of truth can hardly be denied to it."[14]

While the growing consensus among researchers, and the scientific evidence, indicated that new physics was essential to explaining the structure of the atom, Schott quietly continued his efforts to derive radiationless orbits using classical wave theory. This ended up as more of a side project, because his college duties became more intense: in 1923, he became the head of both the departments of applied and pure mathematics, and in 1932 he became the vice-principal of the college.

Finally in 1933, the year that he retired, Schott produced a remarkable breakthrough, in a paper titled "The Electromagnetic Field of a Moving Uniformly and Rigidly Electrified Sphere and Its Radiationless Orbits."[15] He had done what most had considered impossible: found a solution to Maxwell's equations where a distribution of electric charge accelerates without producing radiation.

Schott imagined a hollow sphere, uniformly coated with a thin layer of electric charge, such as a "metal sphere suspended by a fine metal wire in such a manner that it can be earthed or insulated at will."[16] He imagined this sphere set into periodic motion, tracing an arbitrary path over the course of its period: it could move in a circle, a figure eight, or an even more complicated path. With extremely detailed mathematical calculations, Schott showed that such a sphere will not produce any electromagnetic radiation if it oscillates with one of a set of discrete frequencies. These frequencies did not match the stationary states of the hydrogen atom and thus could not be used to model the atom, but Schott had roughly proven what he had set out to do: show that electric charges can oscillate without radiation. These curious structures would eventually be given the name nonradiating sources: they are, paradoxically, sources of radiation which in fact produce no radiation.

But what is the origin of this radiationless motion? Schott himself did not provide any specific explanation, but a hint is implicit in the mathematics of his result. Schott's calculation showed that the moving sphere produces no radiation if the period of its oscillation—the time it takes to make one full round-trip—is a multiple of the time that it takes light to travel the diameter of the sphere. This suggests that the light wave generated at one end of the sphere travels to, and is canceled out by, the light emitted at the other end of the sphere. We again have a phenomenon of wave interference, but a very special form of wave interference. In Young's experiment, the light emitted from two points will destructively interfere at some locations and constructively interfere at other locations. Schott had discovered that it is possible to create a moving distribution of electric charge—a large collection of points emitting light—where the combined waves from all points completely destructively interfere everywhere outside of the region where the charges are moving.

The strangeness of this result cannot be emphasized enough. All of our wireless communications technology, such as radio, cellphones, and Wi-Fi, are based on the principle that oscillating charges create electromagnetic waves. Schott's result suggests that it is possible, for example, to design a cellphone with a certain shape that doesn't broadcast when you make a call: all the energy remains localized in the region of the cellphone itself, trapped by interference.

In science fiction, only one author seems to have come close to using destructive interference as a means of invisibility. In the novel *Slan* (1946), by A. E. van Vogt, the protagonist uses disintegration technology to annihilate any photons that interact with his spaceship, making the ship effectively invisible.

Schott's result had its own restriction. The condition for radiationless orbits that he derived requires that the radius of the sphere be much larger than the length of the path that the sphere follows—the sphere's motion could be said to be more of a "wobble" than an "orbit." The result was nevertheless groundbreaking, and Schott worked for the rest of his life to refine his theory, publishing a series of papers through 1936 and 1937 on the motion of charged spheres.[17]

Schott himself recognized that his model could not explain the Bohr model of the atom, and he instead looked to apply it to other puzzles in fundamental physics. Since Rutherford's discovery of the atomic nucleus, it had been recognized that the nucleus was itself made up of smaller constituent particles. One constituent was the proton, a particle with a positive electric charge exactly the opposite of the electron's charge. Every element of the periodic table represents an atom with a different number of protons. In 1932, British physicist James Chadwick experimentally proved the existence of the neutron, a particle with a mass almost equal to the proton but possessing no electric charge whatsoever. Schott, in his paper on radiationless motions in 1933, suggested that perhaps the neutron might be two of his radiationless orbiting shells, one positive and one negative, orbiting about each other. This idea was, however, quickly discounted by researchers in nuclear theory.

Schott died suddenly in 1937, leaving his studies of radiation unfinished; his final papers were published posthumously. Schott may be considered the last of the great quantum skeptics—one of the final people to challenge the assumptions of the quantum theory before overwhelming experimental evidence of its correctness crushed any reasonable doubt. Schott himself was remembered fondly and respected for his work, as his obituary for the Royal Society indicates: "The mathematical difficulties are enormous, and the skill showed in getting numerical results shows Schott's mastery at its highest. It might be called the supreme attack of a heroic defender before his death. Defeated? Who shall say? I like to think that in future years the work of Schott will always be consulted for inspiration to tackle the difficulties which come across the path of all theories."[18]

Reading Schott's papers in modern times, I am similarly in awe of his ability to manipulate Maxwell's equations to produce beautiful results. Schott was truly an artist in electromagnetic wave theory.

For the next few decades, other researchers would independently rediscover Schott's radiationless orbits and, like him, struggle to find a use for such a compelling result. In 1948, quantum physicists David Bohm and Marvin Weinstein found more general types of radiationless orbits and even showed that it is possible for some of these radiationless charge distributions to self-oscillate without any external forces present.[19] We can imagine one of these self-oscillating systems as a hollow ball filled with water, where the ball is the spherical shell of charge and the water is the electromagnetic energy of the charges. By shaking the ball of water, we can set it into vibrational motion, where the internal motion of the water is balanced by the external motion of the ball. Similarly, Bohm and Weinstein imagined the motion of the shell and the motion of the electromagnetic energy within the shell balancing each other out.

By Bohm and Weinstein's time, the physics of the neutron was quite well established, but new and curious particles had been found that required explanation. In 1947, researchers Cecil Powell, César Lattes, and Giuseppe Occhialini discovered particles called mesons, which had been predicted to exist in 1934 by the Japanese physicist Hideki Yukawa. Bohm and Weinstein took the opportunity to suggest that maybe, just maybe, mesons might be excited states of the common electron, self-oscillating in their own electric fields. This hypothesis was also quickly discounted, as more properties of the meson were discovered that were incompatible with an excited electron model.

In 1964, Professor George Goedecke of New Mexico State University further generalized the work of Bohm and Weinstein, and he boldly suggested that radiationless motions might be used to construct "a 'theory of nature' in which all stable particles (or aggregates) are merely nonradiating charge-current distributions whose mechanical properties are electromagnetic in origin."[20] This hypothesis has also not gained any traction in the scientific community, as modern discoveries in particle physics have shown that the fundamental laws of physics are more complicated—and stranger—than a simple electromagnetic theory can account for.

The preceding examples show that radiationless motions had become, by the mid-twentieth century, a solution in search of a problem. Numerous researchers looked at the remarkable properties of these charge distributions and felt that they must describe *something* in the natural world, and they strove to figure out a way to use them in physics.

At about the same time that Goedecke was struggling with this problem, other researchers were developing new technologies that would make such "invisible" charge distributions important in a very different context.

11
Seeing Inside

In extenuation of Mr. Bland's slight lapse it must be recorded that he had neither the intention nor the inclination to become a skeleton. Such an ambitious undertaking never entered his mind. Bones, in appalling number, were thrust upon him, so to speak. Or, inversely, flesh was removed. In the long run it made little difference how the change occurred. Bland suddenly and confoundingly discovered he had turned to a skeleton. He discovered also that it is the rare individual indeed who regards a skeleton either as a social equal or a desirable companion.

Thorne Smith, Skin and Bones *(1933)*

The discovery of X-rays in 1895 had produced not only awe and wonder but plenty of confusion and panic. The panic was significant enough to motivate journalists to debunk some of the rumored powers of the new rays. For example, the same news article from 1896 that describes "X-ray proof underclothing and other absurdities" leads with the following description of the limitations of X-ray imaging: "The excited public may calm itself. Nobody, not even Mr. Edison, can see its lungs and livers. The most that can be done is to see a shadow of the skeleton of hand or foot—and after all, is this not wonderful enough to satisfy anybody? Even this can be done only under special conditions."[1]

The "Edison" referred to here is the famed inventor Thomas Edison, who had a brief infatuation with X-rays when they were discovered, and he even speculated, without justification, that they might cure blindness.[2] Soon after Röntgen's discovery, Edison assigned his assistant Clarence Madison Dally to work on X-ray technology; by 1902, Dally had developed severe radiation poisoning, and that poisoning led to his eventual death in 1904. It is thought that he is the first person to die from experimentation with radiation. Dally's experience led Edison to swear off X-ray research, and in 1903 he said, "Don't talk to me about X-rays; I am afraid of them."[3]

Limitations and dangers notwithstanding, the public continued to associate X-rays with invisibility well into the twentieth century, and authors of science fiction and popular science alike took advantage of this to get attention. For example, the novel *Skin and Bones* (1933) by Thorne Smith is a comedy that imagines a man inadvertently turning himself into a skeleton by experimenting with the fluoroscopic chemicals that are used in X-ray photography: "It will never be known definitely what chemical combination wrought the amazing change in Mr. Bland's physical composition. Quite possibly the fumes of his strange concoction, together with an overdose of aspirin invigorated by the reaction of much raw liquor, were sufficient to create a fluoroscopic man instead of a fluoroscopic film."[4]

In popular science, the cover of the February 1921 issue of the magazine *Science and Invention* asked the provocative question, "Can we make ourselves invisible?" and included an equally provocative image of a woman seemingly erased by a scientist with a contraption that looks like a pair of cathode ray tubes (fig. 28).

The article was written by Hugo Gernsback, a Luxembourgish American writer, inventor, and magazine publisher, most famous today for his founding of the first science fiction magazine, *Amazing Stories,* in 1926. In his article on invisibility, Gernsback leans heavily

Figure 28. Cover of the February 1921 issue of *Science and Invention.*

into science fiction as well, imagining a future device that would go beyond the limitations of X-rays and allow the imaging of all parts of the human body, tissue as well as bones:

> What will be the practical use of the future machine which the writer has termed "Transparascope" (transparent—shining through; scope—to see)? As our illustration shows, it will be of inestimable value to medicine. It will make it possible for us to see our internal organs in their true colors and in their true shape. For instance, it will be possible for us to watch the heart beat, and the physician instead of listening to the heart beat will be able to see just what is wrong with it. He will be able to examine the lungs, and he need no longer tap the chest to locate the disease. Physicians will be enabled before performing an operation to see for themselves just exactly what is wrong with an organ and no chance operation need be resorted to; no patient need be cut open to find out the real trouble.[5]

Gernsback was eerily prescient in this statement. In a few short decades, new technology would appear that would allow the three-dimensional imaging of the interior of the human body, including organs such as the lungs and liver, and doctors and surgeons would use those images to guide their treatments, just as Gernsback predicted. Gernsback imagined that some new mysterious form of ray would be required to make such imaging possible, but in fact the first researchers in these new forms of medical imaging used the now humble X-ray in ways previously unimagined. The missing ingredient would turn out to be the computer, which allowed data from multiple X-ray photographs to be combined into three-dimensional images of remarkable resolution; the new technique would first be called computerized axial tomography, or a CAT scan, and is now simply called computed tomography.

Gernsback was also remarkably astute in connecting new forms of medical imaging with invisibility. As CAT scans and other imaging technologies were developed, the question of invisible objects would appear in a new and unexpected way.

The story of computed tomography starts with a job resignation. The South African physicist Allan MacLeod Cormack (1924–1998) was working as a lecturer at the University of Cape Town in 1955 when the hospital physicist at the nearby Groote Schuur Hospital resigned. The hospital was using radioactive isotopes for treatment of cancer—a technique known as radiotherapy—and South African law required that a qualified physicist supervise the use of the materials. Cormack agreed to work at the hospital for one and a half days a week for part of 1956.

In the technique of radiotherapy, X-rays are directed toward a target tumor, with the goal of damaging or destroying the tumor; the same harmful effects of X-rays that Edison's assistant Dally suffered from are employed to eradicate cancerous cells. To give an appropriate dose of X-rays to the tumor, however, one ideally must know what tissues they are passing through en route, because X-rays are absorbed in different amounts by different tissues—bones, for example, absorb more than other tissues, which is why ordinary X-ray images show a shadow of a skeleton. In Cormack's time, determining the proper dose was a trial-and-error procedure, based on rough estimates of what the X-rays were encountering on the way to their target. These limitations set Cormack thinking about a new approach: "It occurred to me that in order to improve treatment planning one had to know the distribution of the attenuation coefficient of tissues in the body, and that this distribution had to be found by measurements made external to the body."[6] Cormack started working on the mathematical problem: Suppose one took many X-ray images of the human body, from many directions, and combined that information together—

could one then deduce the full three-dimensional structure of the body? He made rapid progress and by 1957 had done a test experiment to measure the internal structure of a cylinder of aluminum surrounded by an annulus of wood. Unexpectedly, his measurements showed that the very center of the aluminum cylinder had a lower X-ray absorption coefficient than the rest of the structure. The workers at the machine shop that fabricated the cylinder confirmed that the center had been manufactured with a slightly different material; this showed that Cormack's technique could, in fact, discover unknown structures in the interior of target objects.

Over the next six years, Cormack returned intermittently to the problem, having moved to work at Tufts University in Massachusetts in 1957. By 1963, he started working on measuring the structure of metal objects that were not symmetric like a cylinder, and he enlisted an undergraduate to write a computer program to analyze the complicated data. The results clearly showed that he could use multiple X-ray measurements to determine the interior structure of complex objects. Despite this, Cormack recollected that the initial reaction to his achievements was underwhelming: "Publication took place in 1963 and 1964. There was virtually no response. The most interesting request for a reprint came from a Swiss Centre for Avalanche Research. The method would work for deposits of snow on mountains if one could get either the detector or the source into the mountain under the snow!"[7]

As often happens in science, however, others inevitably came around to the same line of thinking, and one of those researchers was uniquely suited for the complexity of the task. Godfrey Hounsfield (1919–2004), an English electrical engineer, learned the basics of electronics and radar while working as a volunteer reservist for the Royal Air Force. In 1949, he took his skills to EMI (Electric and Musical Industries), Ltd., in Middlesex, England, where he continued research into guided weapons systems and radar.[8] This was the era of dramatic advances in computers, and the invention of the transistor in 1947 led to the development of the first transistor computers, replacing the older machines that were based on vacuum tube technology. Hounsfield led EMI's project to develop the first commercially available all-transistor machine made in Great Britain, the EMIDEC 1100.

By the late 1960s, Hounsfield was looking at the problem of using computers for pattern recognition, such as identifying handwriting, fingerprints, and faces. All of these problems were unified in asking: What is the relationship between a complicated set of data collected and the key information we want to extract from the data? While thinking about these problems, he started thinking about X-ray imaging and came to the same conclusion that Cormack had arrived at years earlier: the combined information from multiple X-rays, taken at multiple angles, could provide detailed information about the interior of a structure. He set up some computer simulations of a "black box" containing hidden objects and simulated the X-ray images that would be produced from different illumination directions. With this information, he was able to reconstruct an image of his hidden (simulated) objects. With this encouraging outcome, Hounsfield began to work on a laboratory prototype.

It is worth spending a moment to give at least a simple idea of how this technique works. Let us imagine that, like Hounsfield's simulation, we have a closed box with an unknown object in it and imagine further that the object absorbs all X-rays that hit it. We then shine X-rays from the left and record the brightness of the X-rays on the right (fig. 29). This single image, which is a traditional X-ray, is also known as a shadowgram, as we learn a bit about the object from the shadow it casts. But as the illustration shows, multiple objects could give the same shadow—a square, a rectangle, a circle, or a triangle. We then shine X-rays from below: the narrow shadow that is

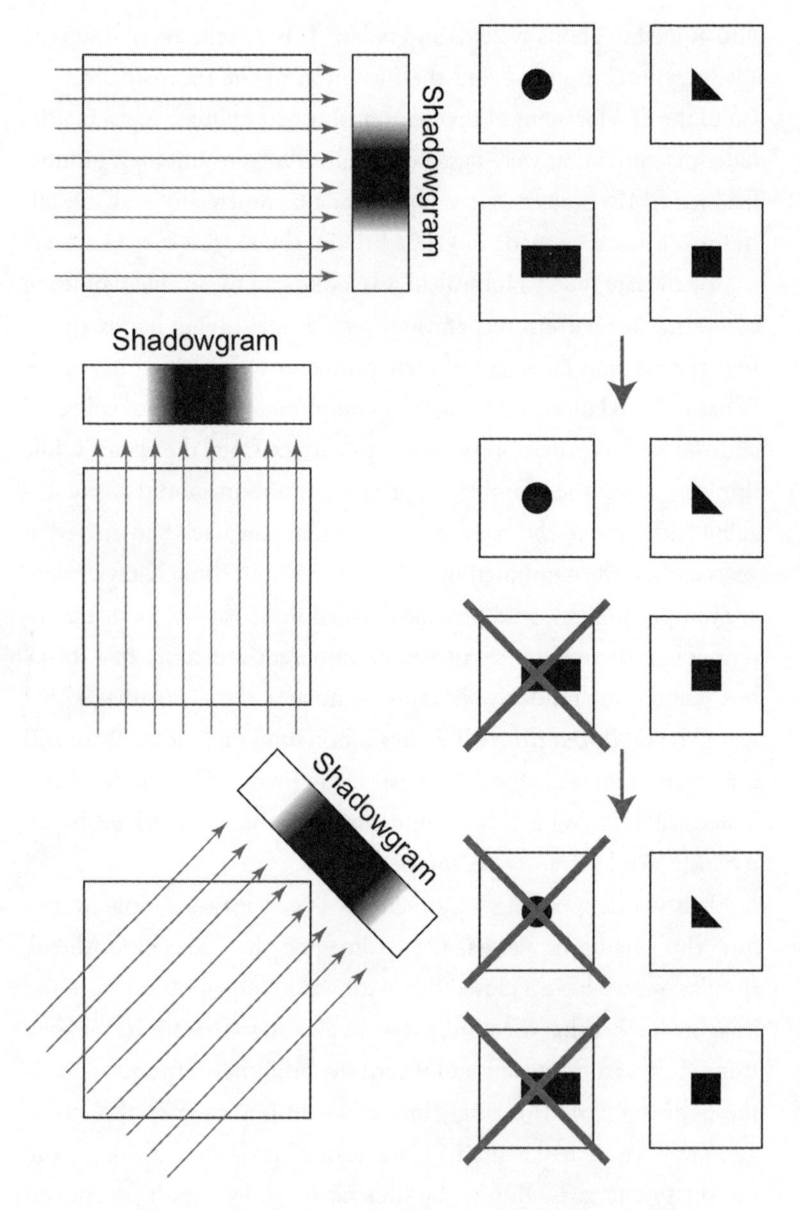

Figure 29. Taking multiple X-ray absorption measurements allows us to learn more and more about the object and narrows down the possibilities for its shape.

cast eliminates the rectangle but leaves open the possibility of the other shapes. Next we shine X-rays from an angle and find that the object casts a wider shadow from that angle, which eliminates the possibility that it is a circle. We still do not know the exact shape of the object, but as we take shadowgrams from more and more directions, we learn more and more information about the object's shape. This is the essence of computed tomography: that information from a large number of shadowgrams can be used to determine quite accurately the shape of the object.

For his first prototype scanner, Hounsfield used a radioactive gamma ray source to produce the imaging radiation. Gamma rays are electromagnetic waves of significantly higher energy than X-rays; presumably Hounsfield used such a source because he had one on hand in the laboratory. He was able to image an object, but the gamma source was of such a low intensity that it took nine days to collect the twenty-eight thousand measurements needed for image reconstruction. Encouraged by positive results, he replaced the gamma source with an X-ray tube and was able to reduce the measurement time to nine hours.

The first tests were done on plastic "phantoms" that crudely represented parts of the human body; when those tests succeeded, he moved on to a preserved specimen of a human brain from a local hospital. The images taken of the brain were excellent, though he soon found that they were too good: the chemicals used to preserve the brain also enhanced the image beyond what would be seen in a living patient. As a more realistic alternative, Hounsfield purchased fresh bull brains and transported them on the subway to his lab; though the images were not as clear as those of the preserved brain, he confirmed that important anatomical detail could still be seen. The measurements were still so slow, however, that the specimen would rot significantly during the process, resulting in degraded images.

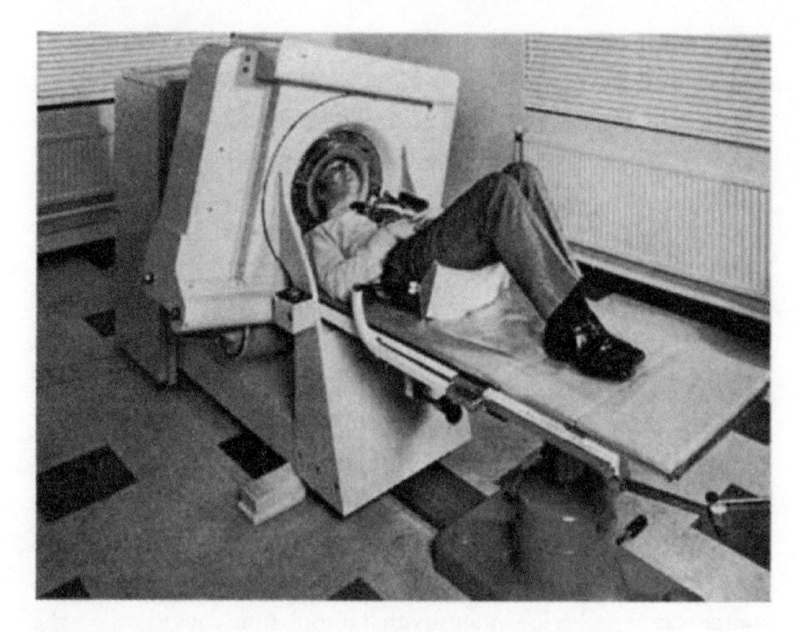

Figure 30. The first clinical X-ray computed tomography machine. Illustration from G. N. Hounsfield, "Computerized Transverse Axial Scanning (Tomography): Part I. Description of System," *British Journal of Radiology* 46 (1973): 1016–223. Republished with permission of British Institute of Radiology; permission conveyed through Copyright Clearance Center, Inc.

All this effort led to the first clinical machine being built and tested in 1972 (fig. 30). The first patient was a woman with a suspected brain lesion, and the computed image clearly showed the cyst in the brain. From that milestone, it was clear that the new technique had real value in detecting and diagnosing illnesses.

In 1973, Hounsfield published his first paper on the technique, which he called "Computerized Transverse Axial Scanning (Tomography)."[9] The term "tomography" comes from the Greek *tomos* (slice) and *graphy* (writing), describing how the machine produces a three-dimensional image of the human body in a series of two-dimensional images that can be stacked on top of each other like layers of a cake.

Hounsfield did not invent the term "tomography": it had been used to describe a different X-ray technique for imaging a slice of a human body in which the X-ray source and detector are moved in circular paths on either side of the patient. This motion causes the produced shadowgram to blur out all parts of the patient except for the slice that lies directly between the source and the detector. Unlike this earlier technique, however, Hounsfield's new tomography was, in essence, exact; it required no blurring and could give a quantitative description of how X-rays are absorbed by different parts of the human body.

Computed tomography almost immediately became a standard tool in medicine; Cormack and Hounsfield shared the 1979 Nobel Prize in physiology or medicine "for the development of computer assisted tomography."

At almost the same time that Hounsfield was developing computed tomography, other researchers were inventing another important medical imaging tool: magnetic resonance imaging (MRI). In this technique, the nuclei of atoms in a patient are excited by magnetic fields, causing them to wiggle in such a way that they produce radio signals. Those radio signals are then measured, and as in tomography, the data are combined in a computer to produce an image of the patient's internal structure. The first MRI scans of a human were done in 1977, and the first clinical scanner was installed in 1980. Over the course of a decade, two new imaging techniques, CAT and MRI, were developed that revolutionized medicine.

The discovery and rapid acceptance of these new imaging methods provided a tremendous boost to a mathematical field of study that had been growing for several decades, known as inverse problems. In most traditional physics problems, one deduces an "effect" from a "cause." For example, when an elementary physics student calculates the trajectory of a ball thrown in the air, they are determining the effect (the path of the ball) from the cause (gravity and the way the ball was thrown). As a more relevant example, when we calculate the electromagnetic radiation produced by a series of oscillating electric charges and currents, we are determining the effect (the measured radiation in all directions, called the radiation pattern) from the cause (oscillating charges and currents).

An inverse problem reverses this procedure and attempts to deduce a "cause" from an "effect." In the case of a thrown ball, the goal would be to use the trajectory of the ball (the effect) to determine exactly how the ball was thrown (the cause). In the radiation example, the inverse problem would be to use the measured pattern of radiation (the effect) to determine the structure of the oscillating charges and currents (the cause). This particular inverse problem is known as the inverse source problem, as one attempts to determine the structure of the source of the radiation. Ordinary vision may be considered an inverse problem, too: our brains interpret the light collected by our eyes as a picture of the world around us.

An early example of inverse problems, before they were even known as such, is the discovery of the planet Neptune in 1846 by John Couch Adams and Urbain Le Verrier. It had been recognized that the orbit of the planet Uranus showed deviations from its expected path, and Adams and Le Verrier independently guessed that the deviations were due to an unknown planet. They did mathematical calculations to predict the position, mass, and motion of this new planet (the cause) from its influence on Uranus (the effect), and the planet was subsequently observed in the area that the calculations indicated.

Because they reverse the usual cause-effect direction of problem solving, inverse problems can suffer from a number of mathematical difficulties. These difficulties can be understood by considering a nonphysics example of an inverse problem: finding a murderer based on evidence left at the crime scene, as in many classic Sherlock Holmes stories. One major difficulty is known mathematically as noncontinuity: small inaccuracies in the data (noise) can lead to a completely wrong solution. In our crime scene, for instance, an innocent person might have shared a drink with the victim hours before the crime, leaving fingerprints at the scene and causing the investigator to mistakenly believe that this person is the guilty party. Because collected data always contain random noise, an inverse problem that is noncontinuous will give nonsense results unless mathematical techniques are used to account for the noncontinuity.

The other major difficulty is nonuniqueness: there may simply not be enough data present to find a unique solution. In our crime scene, this could be a "perfect crime," in which the murderer has not left any evidence to identify him or her, or alternatively there could be so much evidence present, pointing at so many suspects, that it is impossible to choose one over the other. In both cases, the information at the crime scene is insufficient to solve the case. To put it another way, the murderer is effectively *invisible* to our crime scene investigators.

One way to fix the problems of noncontinuity and nonuniqueness is through the use of prior knowledge: information about the solution to the problem that is independent of the data. For instance, in our crime scene analogy, criminal investigators will not base their investigation only on the evidence at the scene but will use interviews of suspects, financial records, and more information to narrow down the list of possible culprits. In imaging problems, prior knowledge can be as simple as knowing roughly the size of the object being imaged: any reconstructed image that is much larger or smaller than expected can be thrown out.

The beginning of inverse problems as a mathematical field of study can be traced back to a paper by the German mathematician Hermann Weyl that was published in 1911.[10] Weyl was, in essence, attempting to answer the question, "Can one hear the shape of a drum?"

Thomas Young and others had noted that the resonant tones of an organ pipe—the frequencies at which it emits sound—depend on the length and diameter of the pipe. Similarly, the frequencies at which a drumhead vibrates depend on the size and shape of the drumhead. Weyl turned this into an inverse problem: If we know the frequencies at which a drumhead vibrates, can we determine the shape of the drum? (Much later, this question would be answered in the negative—the inverse problem would be shown to be nonunique.)

Once quantum physics was discovered and developed, similar questions arose about atoms, which emit light at characteristic frequencies—if we know the frequencies of light emitted by an atom, what can we say about the structure of the atom itself? A key paper on the topic was published by the Soviet-Armenian astronomer Viktor Ambartsumian in 1929, in the German journal *Zeitschrift für Physik,* where it languished without attention for many years. As Ambartsumian later commented, "If an astronomer publishes an article with a mathematical content in a physics journal, then the most likely thing that will happen to it is oblivion."[11] After World War II, however, Ambartsumian's work was rediscovered, and it became a foundational paper in the theory of inverse problems.

Even before the advent of computed tomography and magnetic resonance imaging, many researchers had been studying the aforementioned inverse source problem: the determination of the structure of a source of waves from the measurement of waves. The solution of this problem was deemed important for a variety of different types of waves and applications; a list was provided in 1974 in a report by Norman Bleistein and Norbert Bojarski, reproduced in part below:

- Construction of images of tumors.
- Analysis of subsurface strata for resource identification and recovery.
- Location and identification of discharges in storms for the analysis

of the storms themselves and prediction of tornadoes from characteristic source patterns.
- Location of buried bodies—an important problem in law enforcement.
- Construction of images of airplanes, missiles, surface vessels and submarines.
- Landmine detection.[12]

In all these applications, a source produces radiation that then passes through or reflects off of target objects, which distort the radiated waves. The idea is that by measuring the distortions of the waves, one can determine the object that distorted them, be it a tumor or a buried body.

In that same work, Bleistein and Bojarski also suggested that the inverse source problem could be uniquely solved. However, in 1959 the researcher Harry Moses of New York University had already concluded "that, in general, the sources are not unique."[13] In 1977, the same Norman Bleistein who had argued that the problem was unique now argued the opposite with his colleague Jack Cohen, in a paper titled "Nonuniqueness in the Inverse Source Problem in Acoustics and Electromagnetics"—clearly Bleistein had changed his scientific opinion as new information became available.[14] Bleistein and Cohen further made an important connection: they argued that the nonuniqueness of the inverse source problem was directly connected to the existence of nonradiating sources, of which we have talked extensively before.

In hindsight, this is an obvious connection. The inverse source problem is an attempt to determine the structure of a source from measurements of the radiation it produces. If a source produces no radiation, then it cannot be detected at all and its structure cannot be determined—it is invisible, as far as the inverse source problem is concerned. Clearly an invisible object cannot be detected, but Bleistein and Cohen made an even stronger argument: if nonradiating sources exist *even in principle,* then the inverse source problem is nonunique. A nonradiating source does not have to be present in your measurement in order to make the problem unsolvable; the mere possibility of nonradiating sources proves that the problem is nonunique.

This revelation did not sit well with many researchers who had dedicated significant time and energy to the solution of the inverse source problem. In 1981, Norbert Bojarski published what he believed to be a unique solution, dismissing concerns of nonradiating sources.[15] That same year, researcher W. Ross Stone argued that nonradiating sources simply do not exist.[16] Stone argued, incorrectly, that the only nonradiating sources that are possible are of infinite size, and since no source is of infinite size, nonradiating sources must not exist, and therefore the inverse source problem is uniquely solvable. Stone may not have had access to Ehrenfest's paper from 1911, where Ehrenfest anticipated this argument and showed, with his pulsating sphere, that at least one nonradiating source of finite size exists.

Reacting to Bojarski and Stone, Anthony Devaney and George Sherman of Schlumberger-Doll Research wrote an article on "Nonuniqueness in Inverse Source and Scattering Problems," pointing out in no uncertain terms that nonradiating sources are possible and that their existence definitely makes the inverse problem nonunique.[17] What followed was a particularly heated back and forth, in print, between Stone and Bojarski on the pro-uniqueness side and Devaney and Sherman on the anti-uniqueness side. The frustration can be viscerally felt in the printed reply of Devaney and Sherman to Stone: "First, we would like to draw the reader's attention to Stone's assertion that he has 'presented a *reductio ad absurdum* proof that the fields produced by nonradiating sources are inconsistent with a nonzero source term in the inhomogeneous wave equation.' Stone goes on to say that he has 'purposely' not published this proof. Being acquainted with this so-called 'proof' we can see why he is reluctant to have it published—it is incorrect." Devaney and Sherman concluded with the devastating comment, "If after the material presented in [Reference 12] and the above counterexample, Stone still insists that inverse source problems possess unique solutions, further discussion would be useless."[18]

It should be noted that such heated scientific arguments are not uncommon, and that they are also often part of the scientific process. Each researcher presents their strongest arguments for their case, and it is up to the reading scientists to judge their validity. Devaney and Sherman were particularly frustrated in this case because they had presented quite unambiguous evidence for the existence of nonradiating sources.

Though the argument on nonuniqueness was left unresolved in this printed correspondence, it convinced most researchers that the inverse source problem could not be solved. But this soon raised questions about new forms of three-dimensional computed imaging: How can we be sure that the image that is computed is everything that is really there? For medical imaging, this could very well be an issue of life and death: if a tumor is "invisible," then it cannot be found and treated. As researchers began to introduce new types of imaging beyond computed tomography and MRI, the question of uniqueness and invisibility would be a fundamental one to answer.

12
A Wolf on the Hunt

Nearer and nearer the creature came. It was now sweeping backwards and forwards in its silent beats on our side of the rivulet. Plainly, it was hunting Carl down, and making sure that its victim had not doubled back. Forwards and backwards, backwards and forwards, it passed; and never did I endure moments of more horrible or sickening suspense than whilst awaiting the coming of the beast, whatever it may be, alike unable to defend myself or to fly.

Nearer and nearer it came, and now I could distinguish its shape more clearly. It seemed to me far beyond the common size—but that might be a delusion of the uncertain moonlight or of my overexcited senses—yet it was only a wolf, a solitary wolf!

"Take courage!" I whispered to Carl. "It is nothing but a single wolf. We are two. It will not dare attack us."

"Not dare!" he answered. "Do you know what that wolf is? It is himself! It is Fritz!"

F. Scarlett Potter, "The Were-Wolf of the Grendelwold" (1882)

In 1975, the first scientific paper about a truly invisible object was published in the *Journal of the Optical Society of America.* Written by Milton Kerker and simply titled "Invisible Bodies," it introduced an object that would not scatter any electromagnetic waves illuminating it—the waves would pass through the object and carry on as if they had encountered nothing at all.[1]

This revelation in physics does not seem to have attracted any media attention, though it is understandable why: the objects that Kerker had theoretically described were tiny particles, each much smaller than the wavelength of light and more like a speck of dust, a microscopic water droplet, or a tiny grain of sand. Milton Kerker was the undisputed expert on light scattering by small particles, and so this form of invisibility was a natural extension of his research specialty.

Kerker's invisible bodies are based on the basic physics of light scattering. When light illuminates an object, the electromagnetic waves cause the electrons in the object to oscillate and in the process transfer some of their energy to the electrons. Those electrons, which are now accelerating, produce their own electromagnetic waves, which is the scattered light. Kerker imagined fabricating a small spherical particle that consists of an inner core and an outer shell, floating in a liquid such as water. He then considered the case when both the refractive index of the core is smaller than the refractive index of the liquid and the refractive index of the shell is higher than the refractive index of the liquid. In such a situation, one can show that when the electrons of the core are accelerating upward, the electrons of the shell are accelerating downward, and vice versa. This means that the electromagnetic waves produced by the core and the shell are completely out of phase and cancel each other out by complete destructive interference. With no scattered wave, the illuminating wave passes right through the particle, and the particle cannot be detected: it is invisible (fig. 31).

It should be noted that a single small particle of any type would already be effectively invisible; it would scatter such a small amount of light that it would be unseen by the eye. But when many particles are clustered together, as in a watery mist or a cloud of dust, the combined scattering of all the particles blocks the light from passing

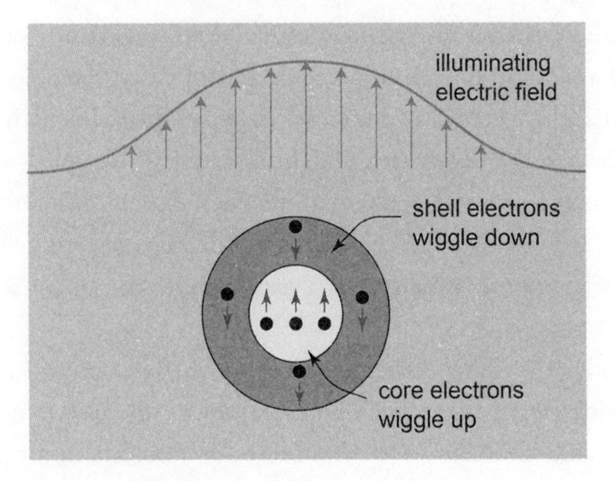

Figure 31. Kerker's invisible bodies and how they work.

through. The same effect causes milk to be opaque: the fat molecules of the milk, floating in water, scatter light. Yet Kerker's invisible bodies, even when put together in bulk, would remain invisible because every individual particle is perfectly invisible by itself. Kerker's research was funded by an unusual source, as acknowledged in the paper: "This research was supported in part by the Paint Research Institute in connection with a numerical study of the effect of microvoids on the hiding power of paints." In essence, Kerker stumbled on invisibility while studying the properties of paint.

Around the same time that Kerker announced his invisible bodies, another researcher was beginning to explore the possibilities of invisibility, in the form of nonradiating sources. This researcher, Professor Emil Wolf, would eventually provide one of the first attempts to prove the existence—or nonexistence—of invisible objects in general. Though a tiny particle can be invisible, could the same be said for larger objects?

Emil Wolf was born in 1922 in Prague, Czechoslovakia, at the beginning of a turbulent era in Europe. As a Jew, he fled the country as a teenager in the late 1930s when Germany occupied the country. From there, he moved through Italy and France. In France, he found work with the Czech government in exile in Paris, and they initially made him a bicycle courier. Wolf found the hectic Paris traffic too much to handle; fortunately, though, the government in exile liked him enough to find other work for him.[2]

When the Germans invaded France, Wolf evacuated by boat to England. By an incredible coincidence, on the boat he found his brother, who had been serving in the Czech army fighting the invaders. Wolf often marveled at this stroke of luck and wondered whether he would have found his brother again after the war if not for this chance meeting; they stayed in contact for the rest of their lives.[3]

In England, Wolf entered Bristol University and earned his bachelor of science degree in 1945. He went on to work toward a graduate degree under Edward Linfoot, a mathematician who specialized in optics. Wolf earned his PhD in mathematics in 1948. At about the same time, Linfoot was appointed assistant director of the Cambridge University Observatory, and he offered Wolf a position as an assistant. Wolf took the offer and spent the next two years working in Cambridge.

During that time, Wolf would often travel to London to attend meetings of the Optical Group of the British Physical Society, usually held at Imperial College. Another regular attendee was Dennis Gabor, the inventor of the three-dimensional image recording technique known as holography. In ordinary photography, one only records the brightness (intensity) of the light on a photographic film, producing a two-dimensional image of a scene. In holography, one records the interference pattern of two waves on the photographic film; one of these is the wave scattered from the object to be imaged, and the other is a "reference" wave. The interference pattern records the phase of the object wave as well as its intensity, and the resulting image then appears three-dimensional, retaining depth and perspective information. Gabor would win the Nobel Prize in physics in 1971 "for his invention and development of the holographic method."

Wolf and Gabor quickly became friends, and Gabor often invited Wolf back to his office to chat about research after meetings of the Physical Society. Through Gabor, Wolf became aware of an extraordinary opportunity to collaborate with one of the great physicists of the twentieth century, the German quantum theorist Max Born. Born was actively involved in the early development of quantum mechanics and is perhaps best known for answering an important question about the waves of quantum particles: What is "waving"? For an electromagnetic wave, it was known that the electric and magnetic fields are the quantities that travel as a wave, but for a quantum particle such as an electron, it was unclear at first how to interpret the electron's wave properties. Born proposed a probabilistic interpretation of the quantum wave function: the wave describes the *likelihood* of finding a particle at a particular point in space. If Young's two-pinhole experiment is done with electrons, the bright spots on the observation screen correspond to locations where the electron is likely to be found, while dark spots correspond to locations where the electron is unlikely to be found. Born would win the Nobel Prize in physics in 1954 for "fundamental research in quantum mechanics, especially in the statistical interpretation of the wave function." It is worth noting that this same statistical interpretation can be applied to the electromagnetic wave properties of a photon.

In 1950, Born was sixty-seven years old and approaching retirement. In 1933, he had released a German book on optics simply titled *Optik,* and he was interested in updating it to include modern developments and publish it in English.[4] Gabor recommended Wolf as an assistant, and Born hired him largely on the strength of Gabor's recommendation. Wolf moved to Edinburgh in January 1951 to help with the book writing. At that time, Wolf was forty years younger than Born; later in Wolf's life, people would often ask if he was the son of the Emil Wolf who had worked on the book with Born. As Wolf recounted, one letter writer cheekily replied in correspondence, when Wolf explained that he was the original author, "Congratulations! You must be 100 years old!"[5]

The book would later be titled *Principles of Optics,* and it took some eight years to finish in the end, much longer than either author intended. During the process, Wolf started doing research on the statistical properties of light, establishing a field of optics now known as coherence theory. All sources of light have some inherent randomness built into them, and this randomness is conveyed into random fluctuations of the emitted light wave. Coherence theory is concerned with the question: How does the random fluctuation of light—its statistical properties—affect the observed and measured properties of light? Coherence theory is, in essence, a mixture of the physics of light and the mathematics of statistics.

In the mid-1950s, Wolf made a major breakthrough; he found that the statistical properties of light travel as a wave analogous to the way light itself travels as a wave. The resulting mathematical equations, now often called the Wolf equations, form the foundation of coherence theory.[6] Wolf met a little resistance to the idea at first: when he explained his results to Born, Born placed his arm on Wolf's shoulder and said, "Wolf, you have always been such a sensible fellow, but now you have become completely crazy!"[7] To Born's credit, he gave Wolf's conclusion further thought and agreed with him a few days later.

Work on *Principles of Optics* was slowed somewhat because Wolf was finishing a chapter on coherence theory for the book, which would make it the first book to contain any such description. When Born learned that this chapter was the holdup, he wrote to Wolf something along the lines of "Who apart from you is interested in partial coherence? Leave that chapter out and send the rest of the manuscript to the printers."[8] Wolf nevertheless finished the chapter and included it in the book, which was published in 1959. In a fortuitous twist of fate, the first working laser was announced in 1960. These new powerful sources of light possess unusual statistical properties, and an understanding of coherence theory became necessary for understanding the behavior of lasers. This made *Principles of Optics* essential reading for almost every optical researcher, and the book has been reprinted and revised numerous times since then; the seventh edition came out in 1999. *Principles of Optics* also contained one of the first textbook descriptions of holography, for which Dennis Gabor was grateful; Gabor later sent Wolf a reprint of a holography paper with the dedication, "I consider you as my Chief Prophet!"[9]

In 1959, Wolf accepted a position at the University of Rochester in New York, where he lived and worked for the rest of his life. He had a remarkably productive career, publishing some five hundred research papers and three major books on optics and coherence. Over his life, he supervised some thirty graduate students, many of whom have become leaders in their respective areas of research.

Wolf worked on many topics of theoretical optics beyond coherence and, like Max Born before him, began studying the scattering of light. In particular, Wolf became interested in the inverse scattering problem: If one shines light on an object from many directions and measures the light scattered from the object in each case, can one determine the structure of the object? In 1969, Wolf combined the scattering theory of Born and the holographic ideas of Gabor and published a paper on "Three-Dimensional Structure Determination of Semi-Transparent Objects from Holographic Data."[10] This work, though it predates Hounsfield's computed tomography by several years, shares strong mathematical similarities with it and later became known as diffraction tomography. Computed tomography, which uses X-rays, ignores the wave properties of light and determines structure entirely through X-ray absorption measurements. Diffraction tomography, as its name implies, uses the wave properties of light, taking into account the diffraction of light by a scattering object.

The study of inverse problems such as diffraction tomography naturally led Wolf to investigate questions of invisibility. At some point, he became familiar with the earlier work on nonradiating sources.[11] In 1973, he published his first paper on the subject, "Radiating and Nonradiating Classical Current Distributions and the Fields They Generate," with his recently graduated student Anthony Devaney.[12] From then on, Wolf performed research on nonradiating sources for much of his career, elucidating their curious and often seemingly paradoxical properties.[13]

It was through nonradiating sources that I became involved with Emil Wolf. Around the third year of my PhD studies, I was looking for a change of research area, and while having lunch at Taco Bell my classmate Scott Carney noted that Wolf was looking for another graduate research assistant. At that point, I was about twenty-five years old and Wolf was about seventy-four—the age gap between us was even greater than the one between Born and Wolf when they started working together. At our first meeting, this was the first thing that Emil brought up. He warned, "You should consider that I'm old and could die at any time, but my doctors say I'm in good health and I'm still doing a lot of work." He then dropped a hefty pile of his published papers on my lap to read, and I realized that all of them had been published in the past five years. Convinced that he was a good choice of adviser, I started working with him immediately.

The research that Wolf suggested to me was a mixture of the theory of nonradiating sources and coherence theory. A nonradiating source, as we have seen, depends on complete destructive interference of the waves emanating from the source; it was not completely clear at that time that a source with random fluctuations—which is called partially coherent—could still be nonradiating. My first published paper on the subject in 1997 showed that even a source with a lot of randomness could be radiationless.[14] From there, my PhD developed into a general study of nonradiating phenomena, which ended up with the title "Nonradiating Sources and the Inverse Source Problem," published in 2001.[15]

My years working with Professor Wolf were some of the most enjoyable of my life. Wolf maintained a vibrant research group, and we would often gather over lunch to debate scientific ideas. Those arguments could get quite heated, sometimes becoming almost shouting matches, but Wolf always stressed that in spite of our disagreements, we would be friends afterward. He truly regarded his students as family members, and they were regularly treated to home-cooked meals and desserts from his wife, Marlies. I had the pleasure of eating many of these meals when I helped Wolf finish the revised index for the seventh edition of *Principles of Optics* (fig. 32).

Emil Wolf maintained many old-school characteristics of professors from his youth. Though we were all friends, he expected his students to call him "Professor Wolf" or "Doctor Wolf," at least until they had reached a stage in their career when they were acknowledged as an equal. For me, this came during a visit to Rochester around 2006 with my girlfriend at the time. We had a lovely Italian dinner, and Beth was chatting with "Professor Wolf." At that point, Wolf said to her, "Please, Professor Wolf is much too formal—call me Emil." He then looked across the table to me, almost as an afterthought, and said, "It's about time—you can call me Emil, too!" So Beth got to call him Emil after about thirty minutes of conversation, and got

Figure 32. Emil Wolf and Greg Gbur celebrating the finished revisions to the seventh edition of *Principles of Optics* in 1999.

to do so before me. Once you were on a first-name basis with Emil, he would never forget and would relentlessly correct you if you dared to call him "Professor Wolf" again.

Part of Emil's interest in nonradiating sources arose from the close relation between the source problem and the scattering problem. In a source problem (also called a radiation problem), a collection of oscillating electric charges produces electromagnetic waves: the "radiation." In a scattering problem, a light wave illuminates a scattering object. That light wave makes the electrons in the scattering object oscillate, which causes them to produce electromagnetic waves: the "scattered light." In the source problem, we assume that electric charges have been set into motion but do not consider the origin of that motion; in the scattering problem, the charges are set into motion by

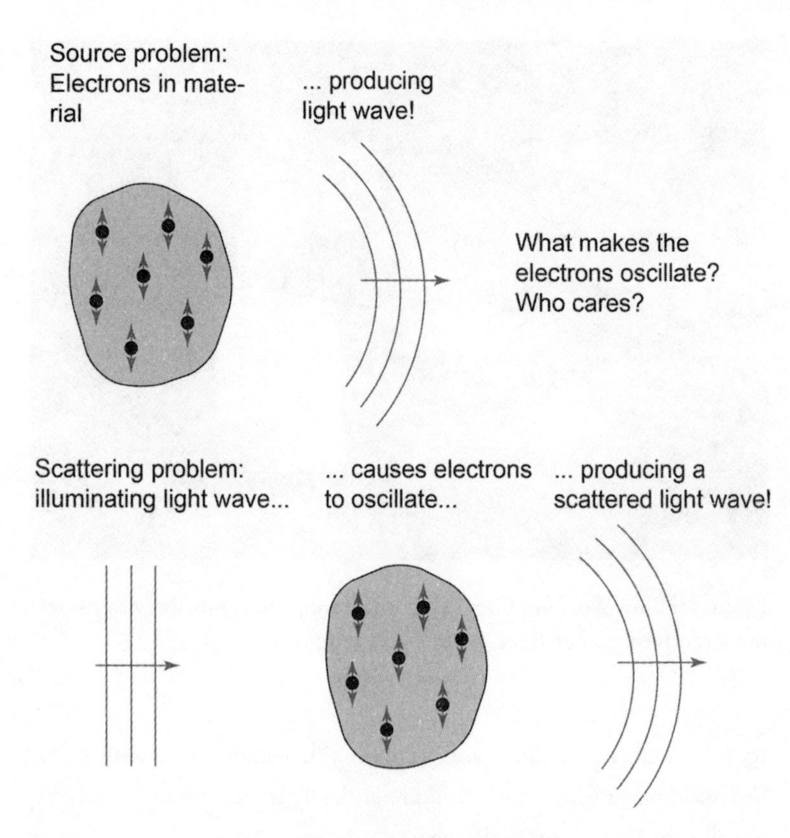

Figure 33. The physics behind the source problem and the scattering problem.

the illuminating light wave. Mathematically, the source problem is equivalent to the scattering problem when we consider an object illuminated only from a single direction (fig. 33).

The existence of nonradiating sources therefore implies that there exist scattering objects that are perfectly invisible when light is shined on them from a single direction of illumination. This in turn means that we cannot solve the inverse scattering problem, and determine the structure of the scattering object, by measuring the scattered field when only one direction of illumination is used. This is in agreement with computed tomography: a single X-ray shadowgram does not give enough information to determine the three-dimensional structure of a patient, and many shadowgrams must be used.

But how many measurements do we need to make—that is, how many directions of illumination do we need to use in order to reconstruct a three-dimensional image of our object? In 1978, Devaney showed theoretically that it is possible to construct an object that is invisible for any *finite* number of directions of illumination, creating a serious concern that the inverse scattering problem might be nonunique.[16] These possible invisible objects were given the very unexciting technical name "nonscattering scatterers," and the existence or nonexistence of nonscattering scatterers became an important question to answer.

The question troubled Emil and his colleagues, but finally in 1993, Wolf and Tarek Habashy showed theoretically, at least for weakly scattering objects, that if scattering measurements are made for an *infinite* number of directions, the inverse scattering problem is unique.[17] A few years earlier, the mathematician Adrian Nachman had provided a general mathematical proof of the uniqueness of the inverse scattering problem for seemingly all scattering objects.[18]

This result may at first glance not be reassuring, since we can never take an infinite number of measurements. But what Wolf, Habashy, and Nachman had shown is that the inverse scattering problem is a process of elimination, similar to computed tomography, described in the previous chapter. As more measurements are made of the scattering object from different directions of illumination, the number of possible structures the scatterer could be becomes smaller and smaller. The structure will never be perfectly known, but with enough measurements we can determine as accurate an approximation to the real structure as we desire.

When I began work on my PhD with Emil around 1996, his work on nonscattering scatterers convinced me that perfect invisibility was impossible. And his theoretical work was correct, but what I failed to realize is that it applied only to a certain class of scattering objects. The class of scattering objects included objects fabricated from almost every material found in nature, but what would happen if one attempted to construct an invisible object out of materials not found in nature? This would be the insight that led to the introduction of invisibility cloaks and the flurry of research that followed.

I continued to work with Emil and visit him even as I became a postdoctoral researcher, and my final paper with him was published in 2004. Emil once described his work with Born as "precious to me beyond description, because it enabled me to see and to talk with him every day."[19] Emil was borrowing the phrase from Born, who had said the same thing about working with Albert Einstein. I in turn look back on my time working with Emil as precious to me beyond description.

13

Materials Not Found in Nature

"Kirk was always the esthete of the family. He thought that the unsightly portions of public structures might be painted with the chemical. Bridges and things. In order to make them more attractive."

"I don't think I understand."

"Well, since sulfaborgonium is an anti-pigment and a total barrier of light rays, it would naturally render these ugly portions invisible. However, I don't think—"

"Wait a minute. Would you go around that corner again, doctor?"

"I beg your pardon?"

"Did you use the word *invisible?*"

Henry Slesar, "The Invisible Man Murder Case" (1958)

In 1890, the German physicist Otto Wiener performed an experiment that would turn out to be a milestone in the study of light. In 1862, James Clerk Maxwell had hypothesized that light is an electromagnetic wave; in 1889, Heinrich Hertz had demonstrated that electromagnetic waves exist, in the form of radio waves. Though there was relatively little doubt that light was an electromagnetic wave at this point, few experiments had been done to confirm this.

Hertz had demonstrated the existence of radio standing waves by reflecting them off of a mirror. Wiener arranged an analogous exper-

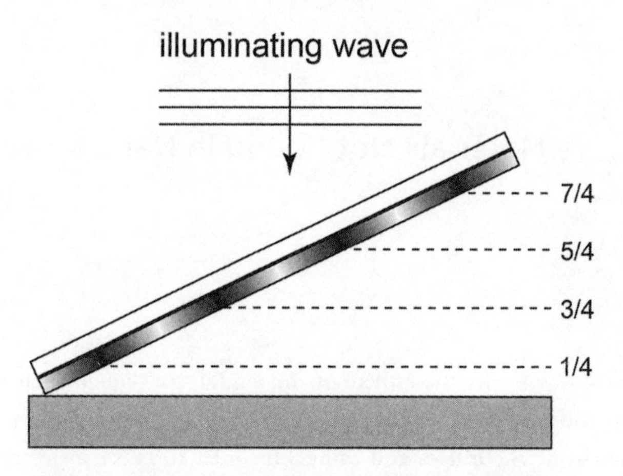

Figure 34. Wiener's experiment: dark spots on the photographic film, indicating that the film has been exposed to light, appear at every quarter wavelength.

iment for light, though he was faced with a challenge: the wavelength of light waves is much smaller than that of radio waves. The wavelength of blue light, for instance, is around a billionth of a meter; an interference experiment with blue light would produce light and dark lines also about a billionth of a meter apart, impossible to observe with the naked eye. Wiener conceived of a simple and clever scheme: he deposited a very thin photographic film on a glass plate and tilted the glass plate slightly with respect to the mirror. The tilt effectively stretched out the interference pattern across the length of the film, making it wide enough to measure (fig. 34).

The goal of this experiment was to test some of the more subtle predictions that arose from Maxwell's equations. Maxwell had predicted that the standing waves of the electric and magnetic fields appear at different spatial locations. The bright spots of the electric field begin one quarter wavelength from the surface of the mirror and then appear every half-wavelength farther away. The bright spots of the magnetic field, however, begin at the surface of the mirror and then every half-wavelength afterward. In performing the experiment, Wiener found that the photographic film was undeveloped right at the surface of the mirror. This indicated that the photographic film had been developed by the electric field of the light wave. He concluded, "In the nodes of the electric forces a minimum takes place, in the antinodes of the same a maximum of the chemical effect; or: the chemical effect of the light wave is attached to the presence of the oscillations of the electric and not the magnetic forces."[1] Wiener had found, almost as an afterthought, that the "active ingredient" of light is the electric field. This would be found to be true for most natural materials, when exposed to visible light: the material reacts to the electric field, not the magnetic field, of the electromagnetic wave.

But a material that also interacts with a magnetic field would behave very differently and could have very beneficial properties. For ordinary materials, reacting to the electric field only, Maxwell's theory predicts that there will almost always be a reflected light wave. However, Maxwell's theory also predicts that a material with just the right combination of electric and magnetic responses could be reflectionless; in essence, the reflected waves created by the electric and magnetic responses of the material can cancel out—the reflected waves will destructively interfere.

The military application of this phenomenon was not lost on scientists and engineers in World War II. The German physicist Arnold Sommerfeld, who is now recognized as one of the great optical scientists, later recalled his work in this area, "During the war the problem arose to find, as a counter measure against allied radar, a largely non-reflecting ('black') surface layer of small thickness. This layer was to be particularly non-reflecting for perpendicular or almost perpendicular incident of the radar wave."[2] Radar was another technology that arose from Heinrich Hertz's discovery of radio waves. *RA*dio *D*etection *A*nd *R*anging uses radio waves to detect objects in a manner similar to the way a bat uses sound echoes: a radar station sends out signals into the air and looks for echoes to return from enemy aircraft.

Sommerfeld and his colleagues were, in essence, trying to find—or fabricate—a new material that would not reflect radar waves, a material that suppresses reflection through the use of a combined electric and magnetic response. It is not clear that they made any significant progress in this goal, but a similar approach was used in the design of the United States B-2 Spirit stealth bomber. The bomber reduces its detectability to radar through two main design choices: its shape and its material. Radar is an active detection system; the radar station sends out a pulse of radar and then looks for a signal bouncing back from the target. An ordinary aircraft, with a rounded hull, tends to reflect radar waves in all directions, guaranteeing that some of them return to the radar station for detection. The stealth bomber has a flat bottom, which means that radar waves primarily reflect in a single direction in accordance with the law of reflection. Since those waves reflect in a single direction, they are less likely to hit a radar station on the ground and be detected.

The material of the stealth bomber is a carbon-graphite composite that absorbs a significant amount of the radar energy incident upon it. With a wingspan of 52.5 meters and a length of 21 meters, the bomber is reported to have a radar cross-section of approximately 0.09 square meters, making it effectively as big as a basketball as far as radar detection is concerned. The modern stealth bomber has largely achieved the radar invisibility that Sommerfeld dreamed of.

This brings us, at last, to the genesis of modern invisibility physics and an entirely new branch of optical science. In the mid-1990s, researchers at the UK-based company GEC-Marconi were also working on techniques for reducing the radar cross-section of structures, and they had developed a carbon-based material that was very effective at absorbing radar. However, they had no idea why their material was so effective. They asked John Pendry, a professor of theoretical physics at Imperial College, London, to see if he could solve the mystery.

On a very small scale, the material consisted of extremely thin carbon fibers that overlapped with each other; the structure is reminiscent of the "forest" of carbon fibers that gives Vantablack paint its extreme blackness. Pendry realized that the unusual radar-absorbing properties of Marconi's material came from this structure.

This observation was a significant revelation. Throughout most of the history of optics, researchers have manipulated the behavior of light in materials through chemistry. By choosing a material with the right chemical properties, one can achieve a desired optical result. But the Marconi material showed that it is also possible to change the optical properties of a material by changing the structure of the material on a subwavelength scale. By making these manipulations of the material structure, it is then in principle possible to design materials with optical properties not found in nature at all.

Pendry and colleagues at GEC-Marconi explored the possibilities over the next few years. In 1996, they showed theoretically that the optical properties of a metal can be dramatically changed if the metal is arranged in a periodic structure of very thin wires; in essence, they showed that they could relocate the interesting optical properties of the metal from the visible light range and make those properties manifest in the infrared light range instead.[3]

Next, Pendry and the Marconi researchers set their sights on the same sort of problem that had plagued Sommerfeld some fifty years earlier: designing a material that possesses a magnetic response as well as an electrical response. By 1999, they published their first theoretical results on the subject, "Magnetism from Conductors and Enhanced Nonlinear Phenomena."[4] In this paper, they introduced for the first time the use of a structure in optics known as a split ring

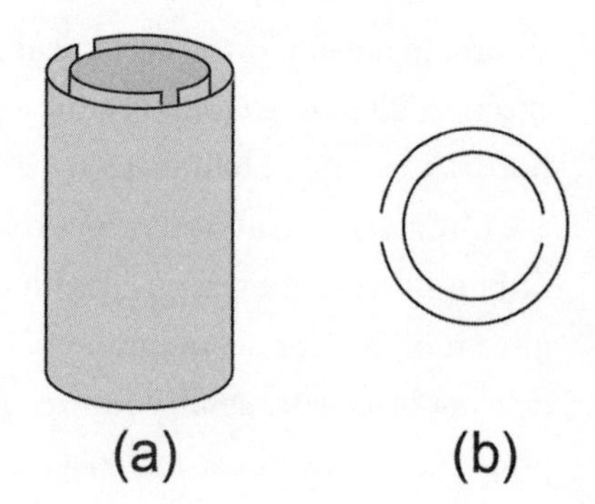

Figure 35. (a) Side view and (b) top view of a split ring resonator.

resonator. This structure is a pair of metallic cylinders nested within each other, with each cylinder possessing a split in it (fig. 35). An entire artificial material would consist of a large number of these structures, subwavelength in size, packed together. These split ring resonators act, in a sense, like organ pipes for light: by an appropriate choice of the cylinder sizes, the gap size, and the metal of the cylinders, it is possible to generate a desired electric and magnetic response at a given frequency, even in principle for visible light, if the resonators can be made small enough. That is, it is possible to make the split rings "resonate" at a desired frequency.

Pendry presented his results on the split ring resonator at the Photonic and Electromagnetic Crystal Structures (PECS) Conference in Laguna Beach, California, in 1999. He also introduced a name for these new artificial materials that have optical properties not found in nature: metamaterials. *Meta* is Greek for "beyond," and therefore the term "metamaterials" refers to materials with properties "beyond" what is found in nature. In particular, Pendry was describing materials whose optical properties come largely from the *structure* of the material, rather than its chemistry.

In attendance at the Laguna Beach meeting were David R. Smith and Sheldon Schultz from the Department of Physics of the Univer-

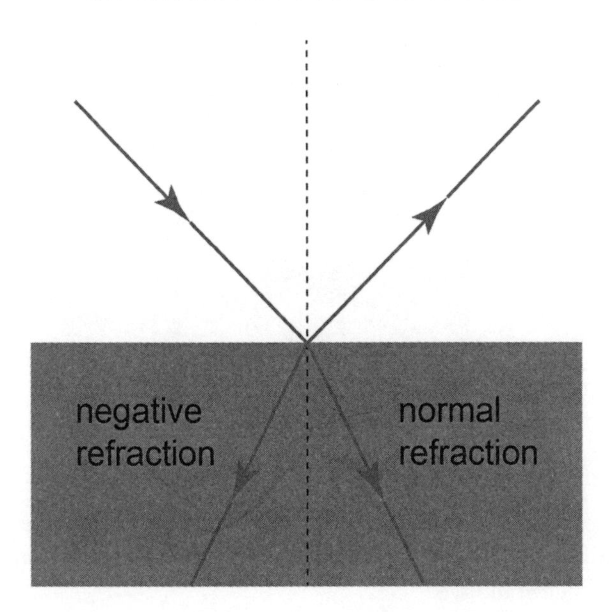

Figure 36. Normal and negative refraction, shown together.

sity of California, San Diego. The physicists noticed a stunning implication of Pendry's work: through the use of split ring resonators, it was not only possible to make a material with a wide variety of magnetic and electric responses but also possible to tune these properties in such a way as to make the refractive index of light a *negative* quantity. For light going from an ordinary medium into a negative medium, light would refract on the *opposite* side of the perpendicular line to the surface (fig. 36). Smith and Schultz talked with Pendry about this and other intriguing optical possibilities at the Laguna Beach meeting, leading to a long-lasting collaboration.

With other colleagues in San Diego, Smith and Schultz wrote a number of papers discussing, theoretically, how to realize a medium with a negative refraction, and in 2001 they published the first experimental demonstration of a material with a negative index of refrac-

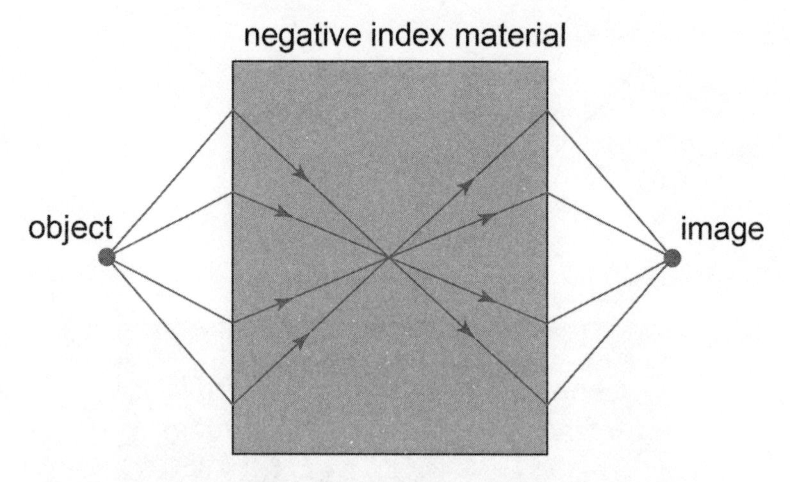

Figure 37. A negative index lens, with refractive index $n = -1$ suspended in air with index $n = 1$. The rays refract twice, forming an image on the opposite side of the slab.

tion.[5] The material was designed for microwaves, with a center wavelength of 3 centimeters; the split ring structures could then be made at a reasonable size of 5 millimeters. (For visible light, with a wavelength of one billionth of a meter, a split ring structure would need to be on the order of one tenth of a billionth of a meter.)

The interest in negative index materials was both practical as well as scientific. In 1967, the Soviet-Russian physicist Victor Vesalago published a theoretical paper arguing that negative refractive index materials are physically possible, and he explained the implications of such materials. Among many intriguing consequences, Vesalago noted that a flat slab of negative refractive index material will act as a lens: unlike ordinary lenses, no curvature of the surfaces is needed to focus light (fig. 37). Vesalago's paper went largely unnoticed when it was published, but the San Diego researchers rediscovered it and used it as a launching point for investigations into negative index materials.[6]

Ordinary lenses have a limited resolution that arises from the wave nature of light. In essence, because a light wave is always spreading, it is usually not possible to concentrate light to a spot smaller than a wavelength. This means that the images of objects that are spaced closer together than a wavelength end up smearing together and cannot be individually distinguished. This had been viewed as a more or less fundamental limitation of optical imaging systems since the dawn of the wave theory of light.

In 2000, however, John Pendry became curious about the resolving properties of Vesalago's flat lens, and he performed some wave optics calculations to determine its resolution. To his astonishment, he found that such a lens is in principle perfect: if it is used to image a pointlike object, it will create a point-like image. At the beginning of the twentieth century, studies of atomic structure were limited by the inability to resolve atoms with ordinary microscopes; it appeared, however, that the Vesalago lens could create a perfect optical image of any object, no matter how small.

Pendry published his result in 2000 in an article provocatively titled "Negative Refraction Makes a Perfect Lens."[7] To say that it caused an uproar in the scientific community would be an understatement; researchers practically fell over each other in a rush to prove that Pendry was wrong. I was still a PhD student at the time, and Emil Wolf received at least one request to be a coauthor on a hurriedly written paper arguing against Pendry's results; he wisely did not take the bait.

Today, everyone agrees that Pendry's calculations are correct. The lens, however, is still not perfect; when one starts to look at practical considerations, such as the absorption of light in the lens, one finds that beyond the resolution limit of traditional lenses there are other limits preventing perfection. The Vesalago lens still cannot be used to image atoms; however, it can provide better resolution than ordinary lenses.

The "superresolution" of the Vesalago lens has been confirmed in several experiments. In 2004, researchers at the University of Toronto designed a superlens for microwaves, confirming that its resolving power was greater than an ordinary lens can achieve.[8] Making a superlens for visible light, with a much smaller wavelength, is significantly more challenging. However, in 2005, researchers at the University of California, Berkeley, showed that a simple thin slab of silver will behave similarly to a Vesalago lens and can provide super-resolution.[9]

For optical frequencies, the challenge is the fabrication of the metamaterials. In order to make a perfect lens or a material with negative refraction, one must be able to control the structure of a material, in three dimensions, on a scale smaller than a wavelength. One can appreciate the difficulty if one imagines the "meta-atoms"—the fundamental units of the metamaterial—as toy building blocks that are a tenth of a billionth of a meter on a side. The task, then, is to assemble these exceedingly tiny blocks perfectly into a structure that has a total length of several centimeters. At this point, we still do not have an efficient and inexpensive method to do this, but it would be a mistake to assume, considering how far the science has come, that it cannot be done.

The announcement of the perfect lens could be said to mark an entirely new era in optical physics. For the entire history of natural inquiry, scientists and natural philosophers had been asking, "What is light?" and "What can light do?" With the introduction of metamaterials, researchers were now asking, "How can we make light do whatever we want it to?"

Many of the rules that optical scientists had labored under for years now turned out to be more like guidelines. This naturally led many researchers to wonder: What else can we do with such materials? One answer, as we now see, was . . . design a cloak of invisibility.

14
Invisibility Cloaks Appear

> The Invisible Weapons Carrier was, in fact, a half-ton of reality. It was large enough inside to contain a man and a fusion bomb, together with the power supply for its engine and its light amplifiers. It bristled with a stiff mat of flexible-plastic light-conducting rods, whose stub ends, clustered together in a tight mosaic pointing outward in every conceivable direction, contrived to bend light around its bulk.
>
> *Algis Budrys, "For Love" (1962)*

The perfect lens of Vesalago and Pendry was in one sense too perfect: the image it produced was exactly the same size as the object being imaged. This made it impractical for applications such as microscopy: in traditional microscopy, the image produced is magnified to a size where it can be recorded or seen with the naked eye. The perfect lens, if used to image an object that cannot be seen with the naked eye, would produce an image that also could not be seen with the naked eye!

Soon after publishing his paper on the perfect lens in 2000, Pendry began looking for ways in which this lens might be modified in order to produce a magnified image. The resulting lens would no longer be perfect, but it would be more practical for use. In 2002 and 2003, Pendry introduced new designs for such magnifying lenses, and

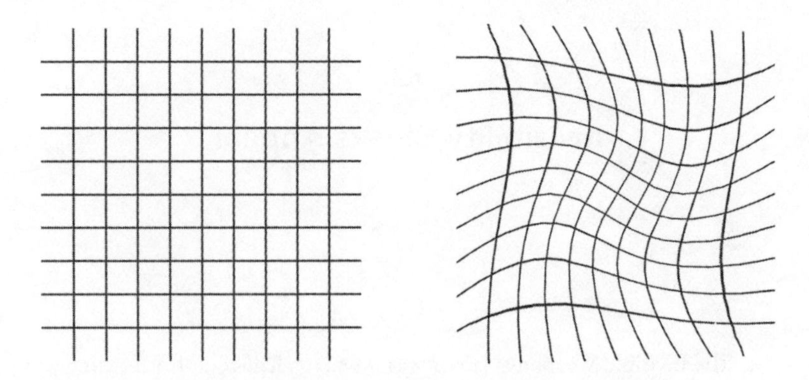

Figure 38. A warping of a discrete mesh for transformation optics. The original mesh is on the left, the warped mesh on the right.

he was aided in the design by a new mathematical tool he had been developing since the mid-1990s that is now known as transformation optics.[1]

As problems in physics and optics have become more complicated, researchers have increasingly turned to computer simulations of the systems in question. For optics, this involves using a computer to solve Maxwell's equations. When light is propagating in a very complicated structure, however, these computations can be difficult to prepare and can take a very long time to evaluate. These calculations are done by representing space on a discrete mesh, where the electromagnetic field is evaluated at the intersections of the mesh. In 1996, Pendry and his student Andrew Ward showed that many calculations can be simplified by performing a mathematical warping of the mesh so that it better matches the actual optical structure to be studied (fig. 38).[2]

In their work, Pendry and Ward noticed a curious property of Maxwell's equations—an optical material can always be found, at least in principal, that will behave like any desired warping of space. This suggested a new strategy: if one wants to design a material that ma-

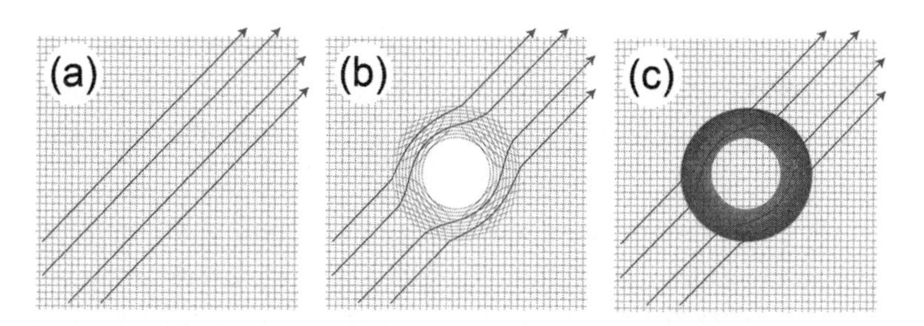

Figure 39. The warping of space that produces an invisibility cloak, and the equivalent material structure that emulates that cloak. Rays of light are shown diagonally passing through the image; their paths are warped just as space is.

nipulates light in a certain way, first determine the warping of space that will produce that manipulation. Once the warping is known, one can use Maxwell's equations to determine the material structure that will produce that optical effect.

Transformation optics uses the same mathematics that Albert Einstein used to develop his general theory of relativity, albeit with a very different interpretation. In Einstein's conception of gravity, a massive object warps the space and time in its local vicinity, which affects the path of other matter and light traveling through space. Transformation optics uses the same mathematics of warped space, but as a tool to design optical devices.

With this in mind, it was almost natural to envision a warping of space that guides light around a central hidden region of space and sends it on its way as if it had encountered nothing at all—an invisibility cloak (fig. 39). We first imagine a region of ordinary space, with the grid showing the relative distance between any two points in space. We then imagine making a tiny puncture in space and stretching that puncture into a hole of finite size. As we stretch that hole, we bend and warp the space around it to make room. Light waves passing through the region are also distorted and pushed out of the central cloaked region. Once we have a mathematical warping that does what we want, we turn to Maxwell's equations to find the equivalent material structure that mimics that warp.

By 2005, Pendry had a good conception of how to use transformation optics to make a cloak. In April of that year, Valerie Browning of DARPA (U.S. Defense Advanced Research Projects Agency) was organizing a meeting in San Antonio, Texas, on metamaterials and asked Pendry to give a presentation. She specifically requested that he "ginger things up," and Pendry decided that a presentation on the possibility of cloaking was just the way to demonstrate the usefulness of transformation optics, a technique that was then basically unknown to the scientific community. As Pendry described the situation,

> At that time I was working on the theory of transformation optics, a very powerful design tool in electromagnetics that I had developed, and thought it would be a good joke to show how to make objects invisible to electromagnetic radiation. My wife suggested that I made reference to someone called Harry Potter, of whom I had never heard but who apparently had something to do with cloaks. However, the joke was taken extremely seriously, and cloaking has since become a major theme in the metamaterials community.[3]

David Smith, now at Duke University, was unable to attend the San Antonio meeting, but he learned of the cloaking proposal and immediately offered to work on an experimental design. His research group therefore had a good head start on building an invisibility cloak before the first official theoretical papers on the subject were published.

As happens quite often in science, however, Pendry was not the only one pondering the possibility of an invisibility cloak. In July 2002, Ulf Leonhardt, then at the University of Saint Andrews, attended a Miniprogram on Quantum Optics at the Kavli Institute for Theoretical Physics in Santa Barbara, California. The discussion topics included metamaterials and negative refraction, and both John Pendry and David Smith were in attendance and gave talks on their work in negative refraction.

Leonhardt had been interested for a number of years in the relation between optics and Einstein's general relativity, and he published his first paper on the subject in 1999, showing how certain moving optical systems can mimic the behavior of a black hole.[4] Listening to the conference discussion on metamaterials, Leonhardt recalled, "I heard talks by David Smith and John Pendry, I heard the arguments for and against negative refraction, and learned about the concept of metamaterials and their experimental demonstrations. It was immediately obvious to me that the next subject in this field is going to be invisibility cloaking."[5] At the Santa Barbara meeting, Leonhardt and Pendry had a number of discussions about physics. Leonhardt shared his knowledge about optical analogues to gravity, and Pendry shared his insights on metamaterials. Neither of them discussed, or was aware, that the other was pondering the possibility of invisibility cloaks.

Leonhardt's technique for designing a cloak was analogous to that of Pendry: create a mathematical warping of space that acts as a cloak and then determine the material structure that will mimic that warping. The principle was straightforward, but making it work was a challenge; Leonhardt struggled with the mathematics of cloaking for the next few years. Finally, on a flight to a workshop in Mexico in September 2005, he figured out the missing piece to finish his cloaking design, and he immediately began work on a paper to publish on his results.[6]

Getting the paper published turned out to be as much of a challenge as solving the cloaking problem. Leonhardt sent it to the prestigious journal *Nature,* where it was rejected, followed by *Nature Physics,* where it was rejected after only two days. He followed up by submitting it to *Science,* another highly prestigious journal, but had it rejected in two weeks.[7]

In early 2006, he sent it to *Physical Review Letters,* long considered the top journal for physics; he met with no better luck there. But one of the reviewers argued that the paper should not be published because the reviewer had attended recent meetings where Pendry had discussed his own (as yet unpublished) work on an invisibility cloak, and thus Leonhardt's work was not new! This was, in fact, an absurd argument for rejecting the paper—unpublished work has no bearing on whether other work should be published. Leonhardt began an appeal with *Physical Review Letters* on these grounds.

But here he would have some unexpected luck. The journal *Science* reached out to Leonhardt. They had now received an article from Pendry, Schurig, and Smith on an invisibility cloak, and the strategy and mathematics for cloaking were so similar in the two papers that the editor opted to publish them together. The two papers appeared side by side in *Science* in May 2006 (fig. 40).[8]

The two papers became an immediate international sensation, and the authors received numerous requests to explain how the devices worked and what they might be used for. The Leonhardt and Pendry teams both used the same analogy to explain how their cloaks operated: "They say it is possible to guide light around the hole, rather like water flowing around a rock in a river, so that the object inside it cannot be seen."[9] This explanation was, curiously, anticipated by one science fiction author decades before. The writer A. Merritt (1884–1943) penned a number of classic novels of science fiction and fantasy and, though he has been largely forgotten today, was wildly

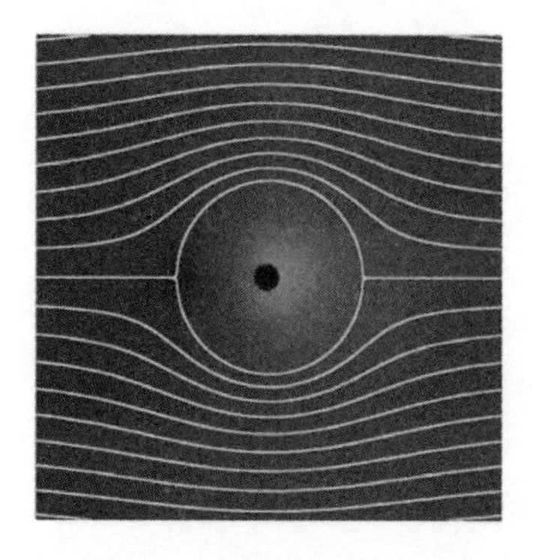

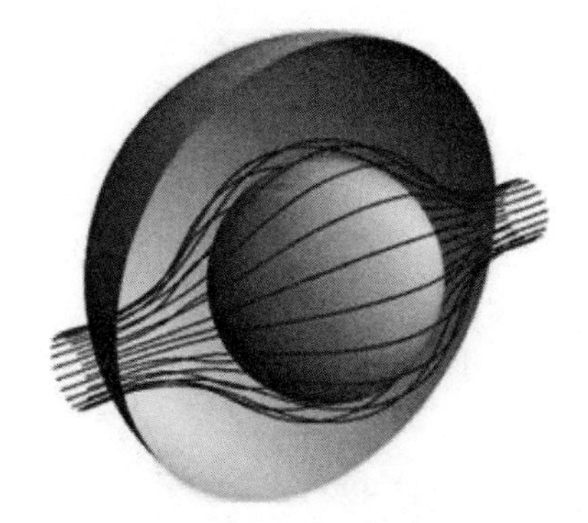

Figure 40. The Leonhardt cloak (left) and the Pendry, Schurig, and Smith cloak (right). Illustrations from U. Leonhardt, "Optical Conformal Mapping," *Science* 312 (2006): 1777–80 (left); and J. B. Pendry, D. Schurig, and D. R. Smith, "Controlling Electromagnetic Fields," *Science* 312 (2006): 1780–82 (right). Republished with permission of the American Association for the Advancement of Science; permission conveyed through Copyright Clearance Center, Inc.

successful in his time. In Merritt's novel *The Face in the Abyss* (1931), an American named Graydon, searching for Incan treasure in South America, instead finds a lost civilization and an evil deity imprisoned in a giant stone face. He also meets the servants of the Snake Mother, winged serpents that can appear and disappear at will. He rationalizes their seemingly supernatural ability as follows:

> The winged serpents—the Messengers? There, indeed, one's feet were solidly on scientific fact. Ambrose Bierce had deduced in his story "The Damned Thing" that there might be such things; H. G. Wells had played with the same idea in his "Invisible Man"; and de Maupassant had worked it out in the haunting tale of the Horla, just before he went insane. Science knew the thing was possible, and scientists the world over were trying to find out the secret to use in the next war. Yes, the invisible Messengers were easily explained. Conceive something that neither absorbs light nor throws

> it back. In such case the light rays stream over that something as water in a swift brook streams over a submerged boulder. The boulder is not visible. Nor would be the thing over which the light rays streamed. The light rays would curve over it, bringing to the eyes of the observer whatever image they carried from behind. The intervening object would be invisible. Because it neither absorbed nor threw back light, it could be nothing else.[10]

Building explicitly on the works of Ambrose Bierce, H. G. Wells, and Guy de Maupassant, A. Merritt managed to accurately imagine how true invisibility might work, at least through analogy!

Other science fiction authors have deduced similar explanations for invisibility. Perhaps the most technically accurate appeared in Algis Budrys's 1962 short story "For Love," in which humanity builds an invisible weapons carrier to deliver a fusion bomb to a hostile alien craft that has taken up residence on Earth. The weapons carrier uses a network of fiber optic cables—a material structure—to guide light around the transport.

When I first read the new invisibility papers of Leonhardt and Pendry and colleagues, I immediately returned to the question, "Why isn't this impossible?" Wolf, Habashy, and Nachman had provided proofs a decade earlier suggesting that true invisibility cannot be achieved.

The two research groups had different answers to this question. I first learned Ulf Leonhardt's answer when we met in 2003 at the meeting in Kiev. When I said, "Perfect invisibility is impossible," he said something like, "Why does it have to be perfect?" Being, for example, 80 percent invisible or 90 percent invisible could still be an incredible advantage. In the movie *Predator* (1987), the imperfectly cloaked Predator can still be detected when it is in motion, yet it is still able to wipe out all of Arnold Schwarzenegger's elite paramilitary team, with the exception of Schwarzenegger himself. In his design, Leonhardt used a simple version of transformation optics that resulted in a cloak that is not perfect but is in principle much easier to manufacture than the Pendry, Schurig, and Smith cloak.[11]

Pendry, Schurig, and Smith's design, however, using an exact version of transformation optics, is in principle "perfect." However, it is not in conflict with the proofs of Wolf, Habashy, and Nachman, because those authors proved that invisibility is impossible for normal materials—and the Pendry, Schurig, and Smith cloak uses metamaterials not found in nature.

In particular, the Wolf, Habashy, and Nachman proofs applied only to nonmagnetic materials. The Pendry, Schurig, and Smith design, however, requires the material of the cloak to have a magnetic response. Furthermore, the cloak must be made of a birefringent material, like the optical calcite that Thomas Young used to argue for the transverse wave nature of light some two hundred years ago. The Wolf, Habashy, and Nachman proofs did not apply to anisotropic materials like calcite.

It should be noted that the physics of the invisibility cloak can also be explained through the use of complete destructive interference, as nonradiating sources and nonscattering scatterers were explained previously. In an invisibility cloak, the scattered field never escapes the region of the cloak, due to destructive interference—we may again imagine that the scattered fields produced by the electric and magnetic components of the cloak end up canceling each other out. One of the many revelations of the invisibility cloak papers, however, was recognizing that transformation optics is a much better technique for designing invisible objects.

In November 2006, only six months after the theoretical papers on cloaking appeared, David Smith and his collaborators at Duke

Figure 41. The first experimental microwave cloak. Photograph courtesy of Professor David R. Smith.

produced the first rough prototype of an invisibility cloak.[12] Like their earlier material that demonstrated a negative refractive index, their cloak was designed to operate at microwave wavelengths and was a flat cloak that was sandwiched between two metal plates (fig. 41). This experimental cloak was of a simplified design that was easier to manufacture and did not provide "perfect" cloaking. However, it did demonstrate that light could be guided around the central cloaked region as the theory predicted.

So in 2006, metamaterials and invisibility cloaks had fully penetrated the consciousness of both the public and the scientific community. There were two big questions on everyone's minds. How could one make an invisibility cloak for visible light—that is, a truly invisible object? And what could these invisibility devices be used for? At that time, it seemed that almost anything could be possible. David Schurig, for example, suggested, "You may wish to put a cloak over the refinery that is blocking your view of the bay," very much the same application Henry Slesar envisioned in 1958 for his invisible paint sulfaborgonium in his short story "The Invisible Man Murder Case."[13] The next decade would feature many attempts to answer both questions, and the answers would often be surprising.

15
Things Get Weird

The case, opened, revealed a space two by three by one-half foot. In it, racked neatly along one side, were twenty little battery cases, with coiled, flexible cables attached, and twenty headsets, bearing curiously complex goggles. The case was practically empty. The Decalon reached in, and with practiced movements passed to her command the goggles and battery cases. Then she reached more carefully into the body of the case. The reaching hand vanished. Presently, queerly section by section, the Decalon was wiped out, till only a pair of feet remained, dwindling off into space. These vanished as some unseen boots were pulled over them.

John W. Campbell, writing as Don A. Stuart,
"Cloak of Aesir" (1939)

The cloaking designs published in 2006 were remarkable in that they in principle provided "perfect" invisibility in the sense that science fiction authors dreamed of. A perfect cloak could hide any object that fits within it, could perfectly guide light around the central cloaked region, and the guided light would be sent along its way without any distortion whatsoever.

Though the papers on cloaking were groundbreaking, everyone recognized that these introductory designs possessed a number of significant limitations. For instance, these cloaks would cast at least a slight shadow. All materials, in their natural state, will absorb light to some extent, even highly transparent materials. At the bottom of a deep pool of water, for instance, everything appears blue because red light has been preferentially absorbed, and deep within the ocean, there is complete darkness because all the light has been absorbed in its journey from the surface. The cloaks described in 2006 are passive optical devices—they guide light around the cloaked region without adding energy to the system—so some of that light will inevitably be absorbed as it is guided. These passive cloaks will cast at least a faint shadow, like the character in Jack London's story "The Shadow and the Flash," making them in principle detectable.

These cloaks also must have extreme variations in refractive index throughout their volume in order to hide the cloaking region, and they must possess extreme anisotropy. Such extreme changes are difficult to manufacture and add to the already difficult and unsolved problem of making metamaterials for visible light.

A third difficulty of the original cloaks: if they were to in fact work perfectly, nobody could see the cloak, but anyone hiding in the cloak would also be unable to see anything! The cloak in principle keeps all light from entering the hidden region, making it a less than ideal solution for espionage. Science fiction authors were certainly aware of this. In the story "Cloak of Aesir," written by John W. Campbell, aliens known as the Sarn have conquered humanity but find themselves opposed by a superpowered being named Aesir with an invulnerable cloak of darkness. In an attempt to trap Aesir, the Sarn deploy their precious invisibility cloaks, which come equipped with goggles that allow wearers to see ultraviolet light. Though the cloaks deflect visible light, ultraviolet light passes through and can be seen, giving the cloak wearer sight. To make invisibility practical, a mechanism for allowing the cloaked object to sense the outside world would need to be included.

Perhaps the greatest difficulty of the original cloaks involves the

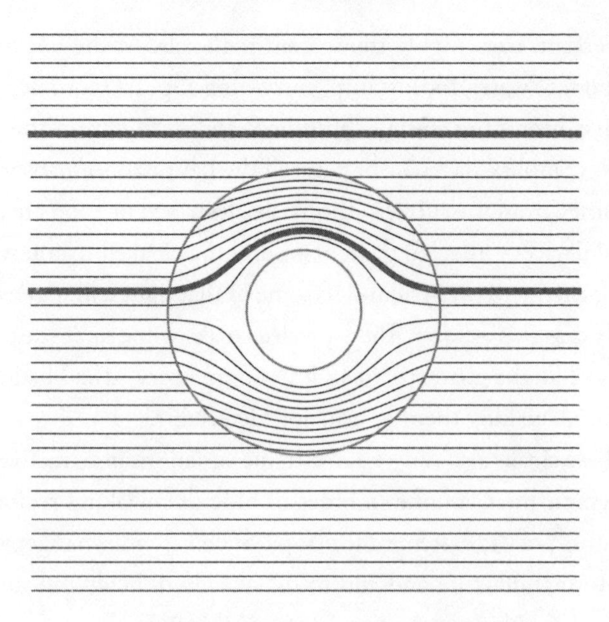

Figure 42. The relativistic limit of cloaking devices. A ray passing near the center must travel a longer distance than a ray traveling outside the cloak.

speed of light. A light ray passing near the interior of the cloak must take a detour around the cloaked region, traveling a longer distance than a parallel light ray traveling outside the cloak. But for the cloak to be truly undetectable, those rays must travel these different distances in the same amount of time: otherwise, one could detect the presence of the cloak by timing how long it takes light to travel through the region. The light detouring through the cloak must therefore travel faster than light traveling in air (fig. 42). But the speed of light in air is almost identical to the speed of light in vacuum, which is the speed limit of the universe: nothing can move faster than the speed of light in vacuum, as Albert Einstein first argued in his special theory of relativity introduced in 1905.

For light traveling in matter, there is a small loophole: for a single wavelength, or a narrow range of wavelengths, the speed of light in matter can be greater than the vacuum speed of light without violating Einstein's relativity, but it has been shown that this range of wavelengths decreases as the size of the cloak increases. Beyond this limited range of wavelengths is an additional problem: other research has shown that cloaks will scatter light more strongly for wavelengths outside the cloaked range. In essence, any cloaking device larger than a microscopic particle could be made invisible for a very specific shade of red or blue but would be perfectly visible or even more visible for any other color.[1]

Because of these limitations, a lot of research on invisibility since 2006 has focused on designing cloaks that are less than perfect but don't suffer from all of the limitations of the original versions. One such design was introduced in 2008 by Jensen Li and John Pendry, and the new approach was referred to as "hiding under the carpet."[2] In the original cloak, transformation optics was used to mathematically stretch a point in space into a three-dimensional cavity, around which light can flow. In a carpet cloak, transformation optics is instead used to stretch a two-dimensional surface upward. If the cloak and the object it is hiding are placed on a flat reflective surface, the light will be redirected as if reflected from the flat surface, hiding the object (fig. 43). The cloaked region looks like the bulge produced by something hidden under a carpet, hence the name "carpet cloak."

Carpet cloaks are a good example of the tradeoffs researchers have been making to achieve practical invisibility. The carpet cloak is not a perfect cloak, in the sense that it can only hide something lying on a flat reflective surface. However, because it does not have to hide the object from all directions, it requires less extreme changes in refractive index and less anisotropy than the original cloaking designs,

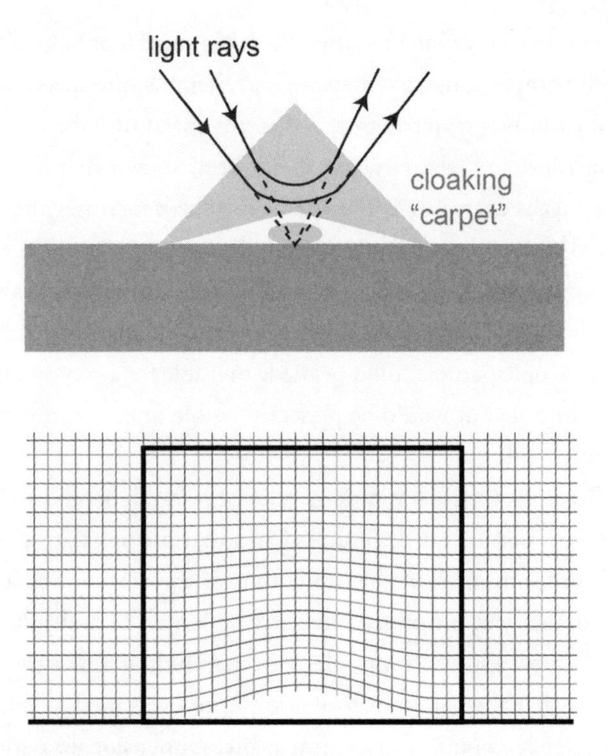

Figure 43. A carpet cloak. Top: it bends rays as if they reflected directly from a flat surface; bottom: the geometric transformation that achieves carpet cloaking.

making it much easier to construct in principle. Because the changes in refractive index are less extreme, the devices are also able to operate over a larger range of wavelengths.

Researchers quickly moved to build prototypes of carpet cloaks. The first of these was constructed by David Smith's group at Duke University and was designed to work for microwaves with a wavelength around 2 millimeters.[3] The cloaked region was a curved bump about 100 millimeters wide and 10 millimeters tall. Not long after, researchers at the University of California, Berkeley, constructed a carpet cloak that could operate for wavelengths closer to visible light, at about 1400 nanometers.[4] The cloaked bump for their device was about 4000 nanometers wide and 400 nanometers tall—a good start, but clearly not anywhere near large enough to hide a person. In 2010, John Pendry collaborated with researchers in Karlsruhe, Germany, to make a device that also worked in a range of wavelengths around 1400 nanometers, with a bump about 1000 nanometers tall.[5]

But what about hiding larger objects? Here, nature provided some unexpected help. Optical calcite—the same anisotropic crystal that led Thomas Young to conclude correctly that light is a transverse wave—has the right anisotropic properties to be used to construct a crude carpet cloak. In fact, a carpet cloak can be fabricated from just two pieces of calcite, appropriately cut and glued together. The cloaks that are created with calcite are far from ideal—they only hide objects from very specific viewing angles and the cloaks themselves are not invisible—but they are relatively easy and inexpensive to create, even at large sizes.

This idea appears to have been hit upon independently, and at almost the same time, by two research groups. In January 2011, a collaboration led by Baile Zhang and George Barbastathis, working in a Singapore-MIT alliance, published the paper "Macroscopic Invisibility Cloak for Visible Light" in the journal *Physical Review Letters.*[6] They constructed a carpet cloak from two pieces of calcite that could hide a region 2 millimeters tall. On a human scale, this still seems exceedingly small, but it was of sufficient size to see the cloaking effect with the naked eye. Furthermore, this carpet cloak worked well for all colors of visible light, making it as close to "true" invisibility as had yet been achieved. Only a month later, researchers from the United Kingdom and Denmark, collaborating with John Pendry, introduced a similar calcite carpet cloak, and the cloaked region also had a height of about 1 millimeter.[7]

The independent publication of two very similar but independent cloaking papers was strikingly reminiscent of the joint publication of the original cloaking papers in 2006, and both cases are an illustration of how ideas in science are often independently discovered when the time is right. A famous example of this phenomenon is the feud that arose between Isaac Newton and the German mathematician Gottfried Wilhelm Leibniz in 1711. Each of them developed the mathematics that is today known as calculus, but they had a fierce public battle over which of them had discovered it first. Newton's supporters went so far as to accuse Leibniz of plagiarizing Newton's ideas; modern historians agree that Newton and Leibniz discovered calculus independently. Similarly, the first calcite carpet cloaks were created independently because all the right knowledge was available for researchers to do so. The two research groups fortunately did not have a feud like Newton and Leibniz.

Baile Zhang, now working at Nanyang Technological University in Singapore, continued to use calcite to create larger versions of carpet cloaks. In 2013, he wowed audiences at the TED Conference in Long Beach, California, with a larger cloak that could hide a bright pink Post-It note. When asked what his plans for the future were, he said that he planned to make the cloak bigger: "As large as possible."[8]

Zhang was not speaking idly. That same year, he and colleagues from China unveiled a modified version of a calcite invisibility cloak that bends light around the sides of an object rather than over the top of it.[9] They made their device large enough to hide goldfish in a fish tank and even a house cat in air (fig. 44).

The original cloak and the carpet cloak were both designed so that light will travel through the inside of the cloak in the same amount of time as it would passing outside the cloak. Zhang's new "ray cloak" dropped this requirement, allowing for a larger device to be built that would operate over a larger range of wavelengths. The tradeoff: as light

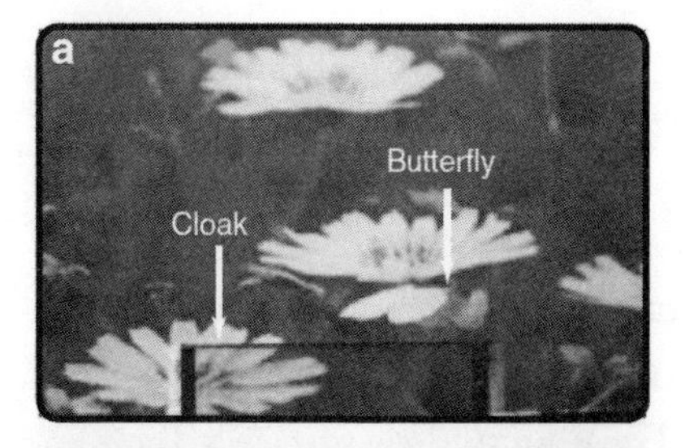

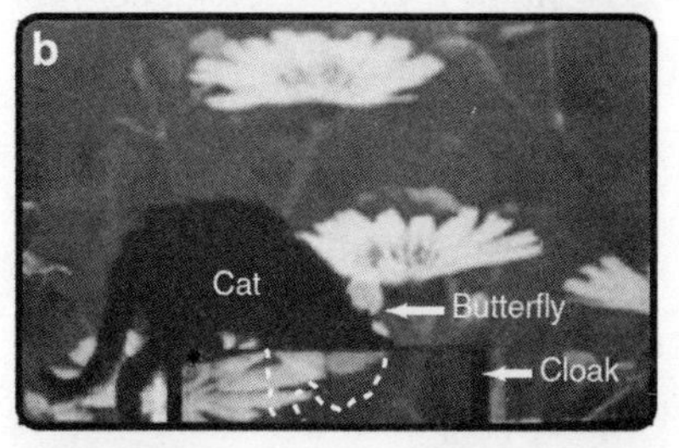

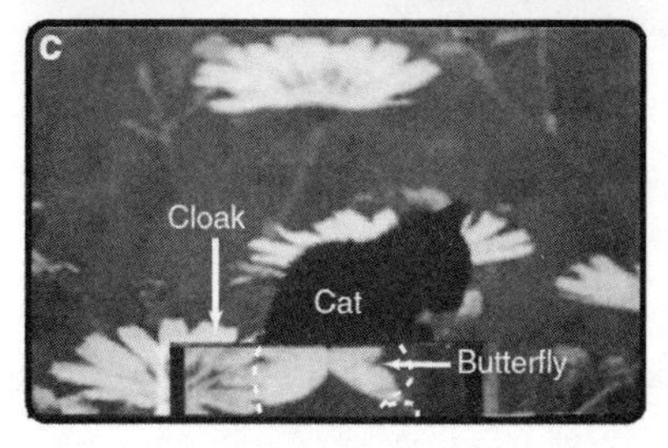

Figure 44. A ray cloak hiding a cat. Illustration from H. Chen, B. Zheng, L. Shen, H. Wang, X. Zhang, N. I. Zheludev, and B. Zhang, "Ray-Optics Cloaking Devices for Large Objects in Incoherent Natural Light," *Nature Communications* 4 (2013): 2652. Licensed under CC BY-NC-SA 3.0.

travels downstream after passing through the cloak, the image will eventually become distorted by this time delay, making the cloak visible. But perfect invisibility was not the goal of Zhang's team; they imagined the cloak could have security and surveillance applications, "where one might imagine hiding an observer in a glass compartment that looks empty."[10] Because such a ray cloak is not invisible for all possible directions of observation, the researchers also suggested that it could be improved by adding an active component: versions of the device could be designed to track the position of observers and constantly orient the invisible side toward them.

With cloaking attracting world attention both in the scientific community and in general, many researchers began exploring other ways to make an object invisible, even revisiting earlier ideas. In 2005, even before the famous cloaking papers were published, Professors Andrea Alù and Nader Engheta of the University of Pennsylvania were looking at the possibility of making spherical objects invisible by applying a thin metamaterial coating to their surface.[11] Their approach, using a multilayered structure to reduce the light scattered by an object, is similar to the microscopic "invisible objects" that Milton Kerker proposed in 1969. Alù and Engheta generalized Kerker's work to consider how a metamaterial coating might make much larger objects invisible.

The invisibility that Alù and Engheta proposed came with a major advantage over the designs from 2006, as they discussed in a paper titled "Cloaking a Sensor" (2009).[12] The structure Alù and Engheta designed does not block light from entering the central region—it uses destructive interference to prevent any scattered light from leaving that region. If a sensor is placed in the middle of the structure, and the structure is specifically designed to block light scattered from the sensor, then the sensor can receive information while remaining largely undetected. Thus the cloaked sensor potentially overcomes the problem of "how does the invisible man see?"

Alù and Engheta's work showed that there are advantages to making a cloak that is tailored to hide a specific object. Another surprising benefit, as demonstrated theoretically by Hong Kong researchers, is that it is then possible to make a cloaking device that does not actually surround the object to be hidden.[13] The cloak in such a case is designed to stand next to the hidden object and includes in its structure an "anti-object" that cancels out all light scattered from the object. This is as peculiar as imagining a magical invisibility cloak that continues to hide you even when you hang it on a coatrack.

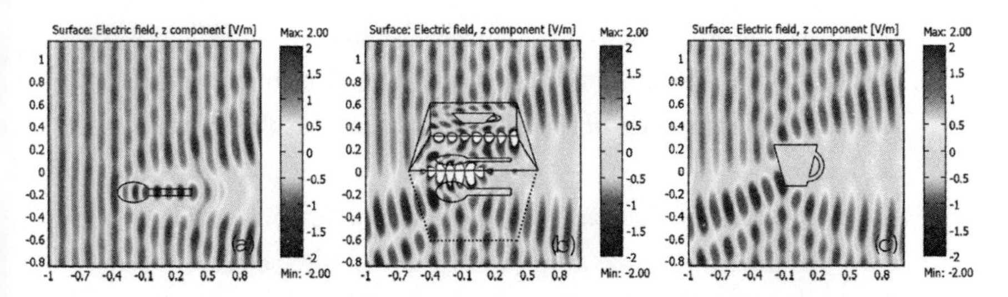

Figure 45. Simulation showing the creation of an illusion making a spoon look like a cup. Illustration reprinted with permission from Y. Lai, J. Ng, H. Y. Chen, D. Z. Han, J. J. Xiao, Z.-Q. Zhang, and C. T. Chan, "Illusion Optics: The Optical Transformation of an Object into Another Object," *Physical Review Letters* 102 (2009): 253902. Copyright 2009 by the American Physical Society.

The introduction of the first invisibility cloaks would lead to even more unusual possibilities. As we have seen, the existence of invisible objects suggests that the inverse scattering problem is nonunique—that is, it is in general not possible to determine the structure of an object from measurements of the scattered field. This in turn suggests that one can use metamaterials to construct an object that looks like any other object—the mere existence of invisibility cloaks indicates that it is possible to create perfect three-dimensional illusions. From my own research on inverse problems, I recognized this possibility the moment the papers came out in 2006 but never had a chance to pursue the idea—one of my big scientific regrets! In 2009, the same researchers from Hong Kong who demonstrated external cloaks performed the first simulations showing the possibility of what they called "illusion optics."[14]

Their first demonstration created an illusion that made a spoon look like a cup (fig. 45). The leftmost image shows the waves, traveling from the left, as scattered by a spoon, and the far right shows the waves as scattered by a cup. The center image shows an illusion device, placed above the spoon, that makes the total scattered light wave look like a cup. The illusion device contains both the image of the cup as well as the anti-image of the spoon, canceling out the scattering of the spoon.

The researchers had an even more dramatic demonstration in store. Because the illusion device can sit next to the object it is transforming, without surrounding it, it is possible to make an illusion device that, when placed next to a solid wall, produces the illusion of a hole in the wall, allowing light to pass freely through the solid wall.

People might be alarmed by the possibility of someone being able to peer directly through their walls, but that concern is premature. The illusion devices, like the original cloaks, work only for a very small range of wavelengths, meaning that a very specific shade of red might be able to pass through the wall, and nothing else. Furthermore, the simulations used a wall that is about one wavelength thick. The wavelength of visible light is about one millionth of a meter; if your house has walls that thin, an illusion device is the least of your problems.

If it is possible to create the illusion of a hole in a wall, is it possible to do the reverse, and create an illusion of a wall in a hole? Also in 2009, a collaboration between researchers in Shanghai and Hong Kong showed that this can in principle be done.[15] Their paper introduced the concept of a "superscatterer," an illusion object that appears much bigger than it actually is. A small superscatterer could then be placed into a large opening, creating the appearance of a solid wall—effectively, an optical secret door.

Other strange devices can be invented by further exploring the analogy between transformation optics and the warping of space in Einstein's general relativity. For example: in Einstein's theory, it is hypothesized that space and time can be fashioned into "wormholes," tunnels that connect distant regions of spacetime. They are named wormholes because they are like the hole a worm tunnels through an apple, providing a shortcut from one side of the apple to the other. In 2007, Allan Greenleaf, Yaroslav Kurylev, Matti Lassas, and Gunther Uhlmann theoretically demonstrated the possibility of creating optical wormholes using transformation optics.[16] These optical wormholes are not true tunnels through space, as in general relativity, but rather passages that allow light waves to pass through them while leaving the rest of the structure invisible.

Greenleaf, Lassas, and Uhlmann were well positioned to make an impact on the physics of invisibility. In 2003, several years before the seminal cloaking papers appeared, the trio published a paper showing that anisotropic materials make a certain class of inverse problems nonunique and lead potentially to invisible objects.[17] This work showed that there was a potential loophole in the nonexistence proofs of Nachman, Habashy, and Wolf, though this would be recognized only in hindsight.

Though an optical wormhole seems like a science fiction dream of the far future, the principles have already been demonstrated in a limited sense. In 2015, researchers at the Universitat Autònoma de Barcelona in Spain constructed a device that acts like a wormhole for static magnetic fields, such as those produced by a bar magnet.[18] If the North pole of a bar magnet is placed in the device, it effectively stretches out the magnet, making the North pole appear to emerge from the far side of the structure. The device uses a combination of magnetic metamaterials and superconductors to achieve the effect.

The creation of magnet-stretching metamaterials has one interesting implication related to fundamental physics. All magnets known in nature possess a North and a South pole, and isolated North poles and South poles are never seen. If one breaks a bar magnet in half in an attempt to separate the poles, one instead gets two bar magnets, each with its own North and South poles.[19] In 1931, the famed quantum physicist P. A. M. Dirac asked theoretically: What would happen if an isolated magnetic pole, a monopole, existed?[20] Using a combination of Maxwell's equations and quantum physics, Dirac showed that the mere existence of a single magnetic monopole in the universe would require all electric charge to come in discrete, quantized bits. Of course, we already know that electric charge comes in discrete amounts, with the smallest electric charge being carried by the electron. This led many researchers to suspect that magnetic monopoles must exist, and there have been extensive experimental searches to look for such monopoles.[21]

In Dirac's calculation, he creates a monopole by mathematically "stretching" a bar magnet, leaving one pole fixed and stretching the other one infinitely far away. The magnetic wormhole theorized by Greenleaf and collaborators and the one experimentally built by the Barcelona group essentially do the same thing: if one end of a bar magnet is placed in the wormhole, it gets stretched—or its fields get stretched—to the other end of the wormhole. Though this demonstration does not prove the existence of true monopoles in nature, it shows that the idea of a monopole is not as far-fetched as it might appear.

One other curious result related to cloaking is worth noting. In 2008, scientists from Shanghai and Hong Kong demonstrated that it is possible to make an "anti-cloak."[22] Researcher C. T. Chan and colleagues showed through simulations that their anti-cloak, when placed inside the hidden region of a cloaking device, would cancel out the effects of the cloak and make the entire structure visible. This result demonstrated that even a "perfect" cloaking device has limitations and can be broken under unexpected circumstances.

As time has passed, the initial furor over cloaking and invisibility has died down somewhat. Researchers still do not have a method for constructing the three-dimensional metamaterials that would be needed to realize the strange science fiction–like devices mentioned in this chapter. Furthermore, most of the devices mentioned suffer from the limitation that they can only work for a very small range of wavelengths.

However, that latter limitation might change in the future. In 2019, an international research collaboration led by Swiss researcher Hatis Altug published a paper in the prestigious journal *Nature Communications* with the provocative title "Ultrabroadband 3D Invisibility with Fast-Light Cloaks."[23] Most of the devices that we have talked about throughout this book have been *passive* devices, which only guide light around a cloaked region without adding energy to the light wave. The Altug cloak is, in contrast, an *active* device: the atoms of the cloak are primed to store energy and can give their energy to an incident light wave as it passes. This allows the light wave traveling through the cloak to appear to move faster than the vacuum speed of light, overcoming the wavelength limitations of the passive cloak. With this active invisibility approach, the researchers simulated a cloak that works for the entire visible light spectrum.

How can light traveling through an active medium appear to move faster than the vacuum speed of light without violating Einstein's relativity? Let us imagine a bus carrying a group of passengers in its center. At each stop, a new passenger gets on at the front of the bus, and a current passenger gets off at the back of the bus. If the new passengers stay near the front of the bus, the mass of passengers has effectively moved forward in the bus. If we were to calculate the center of mass of the passengers, we would find that the passenger mass is moving slightly faster than the bus itself!

Analogously, we imagine that our bus represents a pulse of light traveling through our active medium. As our pulse encounters atoms, it picks up some of the energy of those atoms (passengers entering at the front of the bus) and deposits energy back into the atoms it has already depleted (passengers leaving at the back of the bus). Even though the pulse itself (the bus) is not moving any faster than before, the energy in the pulse shifts forward, giving it the appearance of moving faster than the vacuum speed of light.

If appropriate materials are chosen for the cloak, it can then operate over a wide range of frequencies. The cloaking will still not be perfect—since nothing can travel faster than the vacuum speed of light, a very short pulse will still show a time delay between light passing in the cloak versus outside the cloak—but under ordinary viewing conditions, it could be virtually undetectable.

Don't expect invisibility cloaks anytime soon, however. The Altug paper contained only simulations, not experiments, and the challenges involved in actually fabricating such a three-dimensional metamaterial cloak of practical size remain enormous. The work done by Altug and collaborators shows, though, that the physics of invisibility likely still has some surprises in store for us.

16

More Than Hiding

Suddenly, in his stooping position, a kind of giddiness seized him, and the earth seemed to pitch deliriously beneath his feet. He staggered, he lost his balance, and fell forward from the verge.

Half-fainting, he closed his eyes against the hurtling descent and the crash twenty feet below. Instantly, it seemed, he struck bottom. Amazed and incomprehending, he found that he was lying at full length, prone on his stomach in mid-air, upborne by a hard, flat, invisible substance. His outflung hands encountered an obstruction, cool as ice and smooth as marble; and the chill of it smote through his clothing as he lay gazing down into the gulf. Wrenched from his grasp by the fall, his rifle hung beside him.

He heard the startled cry of Langley, and then realized that the latter had seized him by the ankles and was drawing him back to the precipice. He felt the unseen surface slide beneath him, level as a concrete pavement, glib as glass. Then Langley was helping him to his feet. Both, for the nonce, had forgotten their misunderstanding.

"Say, am I bughouse?" cried Langley. "I thought you were a goner when you fell. What have we stumbled on, anyhow?"

"Stumbled is good," said Furnham reflectively as he tried to collect himself. "That basin is floored with something solid, but transparent as air—something unknown to geologists or chemists. God knows what it is, or where it came from or who put it there."

Clark Ashton Smith, "The Invisible City" (1932)

We usually think of physics as something that is done in a lab, with careful, repeated experiments leading to new insights into nature. But sometimes major discoveries are made in the most unusual and unexpected of places.

Such was the case in 1995, when a singular event occurred in the North Sea, about 160 kilometers southwest of the southern tip of Norway. In 1984, a natural gas platform known as the Draupner platform was built at that location, and it is a major hub for monitoring the flow of natural gas from Norway's various offshore pipelines. The first platform built was a crewed installation known as Draupner S; in 1994, a crewless platform known as Draupner E was added, with the two connected by a metal bridge.

Draupner E was itself a bit of an experiment: it was the first major offshore platform to use a bucket foundation, in which the base is supported by steel buckets sucked into the ground. Because of the use of novel supports, the platform was outfitted with an extensive array of sensors that would continuously measure its motion and the forces it experiences, to confirm that the structure remained stable, especially during the intense winter storms of the North Sea.

On New Year's Day 1995, Draupner E would get a more severe test than anyone ever imagined. At 3:00 p.m., in the midst of a storm, the platform was hit with a monster wave of a size well beyond what was even thought possible. A laser rangefinder on the platform measured the height of the wave at 25.6 meters, or 84 feet; the significant wave height during the storm was only 12 meters (39 feet), or less than half of the monster. The immense size and power of the wave was recorded on other measuring devices, which confirmed that a tremendous, unprecedented force had hit the platform. Fortunately, the platform itself suffered only minor damage from the watery assault, and the crew were safely inside of Draupner S at the time and not even aware that something significant had happened.

For centuries, sailors had spoken of monster waves, appearing out of nowhere, even on relatively calm days: literal walls of water that had the potential to smash ships to pieces. These reports were often dismissed; ironically, as the scientific understanding of water wave physics increased, conventional theories suggested that such waves, though possible, should be so unlikely to occur as to never be seen. The Draupner wave revolutionized wave physics and sent scientists scrambling to understand the origins of what are now known as rogue waves or freak waves, essentially founding an entirely new area of wave research.

Today, it is accepted that a number of factors can contribute to the creation of rogue waves. The focusing of water waves, like light being focused in a lens, can increase the wave height.[1] Regions where opposing ocean currents collide can contribute to the amplification of wave heights. An example of this is at the southern tip of Africa, where the westgoing Agulhas Current bumps into eastgoing currents from the Atlantic—this is also an area where rogue waves are now known to appear with some regularity. The greatest contribution, however, is likely what is known as a nonlinear effect: in short, circumstances can develop where a larger wave "eats" a collection of smaller waves, growing in size with each "bite" until it is truly monstrous.

Rogue waves are now recognized as a genuine threat to ocean transportation and the safety of those on ships; it is believed that more than twenty-two supercarriers have been lost between the years 1969 and 1994 due to such waves.[2] And even without taking into account rogue waves, ordinary waves damage offshore platforms and buoys, and any strategy to minimize this damage would be a boon.

In 2012, Mohammad-Reza Alam of the University of California,

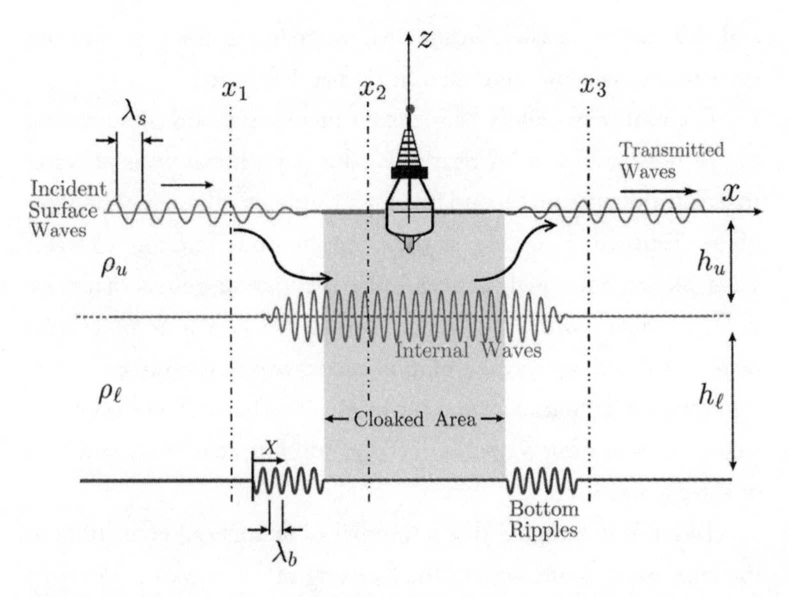

Figure 46. The concept of sea cloaking. Illustration reprinted with permission from M.-R. Alam, "Broadband Cloaking in Stratified Seas," *Physical Review Letters* 108 (2012): 084502. Copyright 2012 by the American Physical Society.

Berkeley, proposed a novel solution: an invisibility cloak for water waves (fig. 46). Alam's strategy takes advantage of the existence of internal waves that can develop below the surface of the water. These internal waves can be generated at depths where the water density changes dramatically from one constant value to another.

If the water depth is sufficiently shallow, surface waves can interact with a structure on the ocean floor—we may call it an "ocean metamaterial"—and such structures can be designed to couple a surface wave into an internal wave, and vice-versa. The result is that damaging waves approaching the cloaked region will be transported *below* the region, leaving the surface relatively calm.

The results discussed by Alam are theoretical and computational only, but they highlight the fact that cloaking can be used for more than simply making something invisible—they can be used to protect an object. As Alam himself states, "We note that, specific to ocean applications, the cloaking is more important in protecting ocean objects against powerful incoming waves than making their trace invisible."[3]

The future of cloaking technology may therefore involve more than just trying to hide an object. It also has potential as a way to protect objects from being damaged by other types of waves. Protection is a much more modest and achievable goal than invisibility. An invisibility device that is 50 percent invisible would be a pretty ineffective way of hiding, but an ocean cloak that repels 50 percent of damaging waves might make the difference between the survival or destruction of the protected object. As we conclude our discussion of the history of invisibility, we look ahead to how invisibility technology might be used for applications beyond hiding things.[4]

To begin, we note that the mathematics of transformation optics, key to the initial designs of invisibility cloaks, can be used to design other novel optical devices. For example, in 2008, researchers Do-Hoon Kwon and Douglas H. Werner demonstrated that transformation optics can be used to design fiber optic cables possessing sharp ninety-degree bends.[5] Fiber optic cables form a key part of our communications infrastructure and are, at their core, simply long, thin transparent glass cables through which information can be transmitted via pulses of light. When sharply bent, however, these cables will "leak" light, causing a loss of signal. Kwon and Werner showed that it is possible to use transformation optics to design a new type of bend in an optical fiber that will result in no light leakage. Such bends could be used to design fiber optic systems that take up less space.

We can imagine how this design process works—part of the power of transformation optics is being able to invent new devices just by

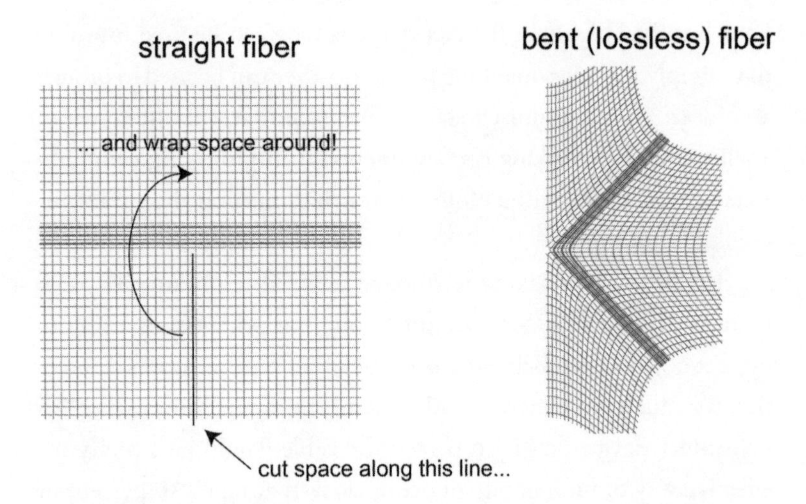

Figure 47. How transformation optics might be used to design a fiber bend. Author's own work.

using a little imagination. We again take flat space, this time with an ordinary straight optical fiber running through it, and mathematically cut space and wrap the left side of the cut clockwise around the center of the region (fig. 47). The result is a region with a ninety-degree bend in it; we may then determine the type of material structure that will produce the equivalent warping of space.

The same authors also used transformation optics to design a wave collimator, which will collect light waves traveling in different directions and send them all traveling in the same direction, and in addition they designed a new type of flat lens to focus light waves. That same year, the familiar collaboration of researchers from Duke University and Imperial College demonstrated how transformation optics could be used to make perfect beamsplitters, which take a beam of light and divide it equally into two, without any light being reflected. In 2012, researchers at Vanderbilt University showed that transformation optics can be used to design devices that can couple light from optical fibers to silicon microchips with minimal loss. In the future, we may see transformation optics transform our communications technology.[6]

The most intriguing use of transformation optics, however, has been the study of cloaking for other types of waves and fields. Light is an electromagnetic wave, meaning that it is a combination of electric and magnetic fields, oscillating rapidly together. Soon after the introduction of invisibility cloaks for light, researchers began studying whether it is possible to create cloaks that will hide objects from *static* electric and magnetic fields—that is, fields that do not change in time. This would include, for example, the electric field produced by powerful electric capacitors as well as the magnetic fields produced by powerful permanent magnets.

In 2007, Ben Wood and John Pendry designed the first metamaterials that could work for static magnetic fields. The design requires the use of superconductors, materials that have zero electrical resistance, fashioned into a series of plates that are then assembled into a magnet cloak. In the following year, Wood, Pendry, and collaborators demonstrated that these superconducting metamaterials work as theoretically predicted, though they did not go so far as to build a cloak.[7]

In 2009, additional simulations of magnetic metamaterials composed of thin superconducting plates were done by researchers in Spain; these were the same researchers who would later construct the "magnetic wormhole." In 2012, the Spanish researchers collaborated with a Slovakian team to introduce a new, simpler design for a magnetic cloak, and they demonstrated experimentally that it worked as predicted. The technique is beautiful in its simplicity, consisting of a ferromagnetic ring of iron surrounding a superconducting ring. The iron ring pulls magnetic field lines into it, while the superconducting ring repels magnetic field lines. If the thickness of the iron ring is appropriately chosen, the combined system will repel magnetic field lines from the cloaked region while producing no distortion of the magnetic field outside the device.[8]

Also in 2012, researchers from Lanzhou and Nanjing in China experimentally demonstrated a cloaking device that protects against static electric fields.[9]

What might be the applications of such static field cloaks? Modern computer systems can be very sensitive to damage from static electricity discharges; electronics that must be placed near strong electrical devices could be shielded by the use of a static electricity cloak. Powerful magnets are used in magnetic resonance imaging and produce such strong forces in their vicinity that operators must take care not to bring any ferrous metal near them. A magnetic cloak could be used to position electronic devices closer to the MRI system.

Transformation optics has been used to design even stranger types of cloaks. Again in 2012, French researchers introduced the concept of "transformation thermodynamics," in which one uses transformation techniques to control the flow of heat rather than waves.[10] Ordinary heat, which is the manifestation of the random motion of molecules, does not travel like a wave but rather diffuses through a region. It is impossible to keep heat out of a region indefinitely; a cold flask on a hot day will, given enough time, warm up to match the outside temperature. But using transformation thermodynamics, the researchers showed that they could slow the flow of heat into the cloaked region. A thermal cloak, combined with traditional thermal insulation, could provide superior resistance to heat.

Cloaks for sound waves have also been introduced and simulated. In 2008, an international collaboration studied such a three-dimensional acoustic cloaking shell. Acoustic cloaking is not burdened by the same relativistic limitations as optical cloaks, making the implementation of broadband acoustic cloaks much more feasible; in 2011, researchers at the University of Illinois, Urbana-Champaign, successfully designed and experimentally tested an acoustic cloak for ultrasonic waves.[11]

People typically associate the term "acoustic wave" with sound waves propagating in air, but acoustic waves can also propagate in solid matter. Research into acoustic cloaking led naturally into an even more intriguing possibility: seismic cloaking, or protection from earthquake waves. An early mention of such a possibility appeared in a paper by researchers from Marseille, France, and Liverpool, United Kingdom, in the journal *Physical Review Letters*.[12] The researchers computationally demonstrated the ability to cloak a region embedded in a thin, flat plate from vibrations excited in the plate. They suggested several possible applications of this technology: to protect delicate parts of automobiles from constant road vibrations and, on a much larger scale, to protect structures from damaging earthquake waves.

Earthquake waves are much more complicated than electromagnetic waves and come in four broad types: P-waves, S-waves, Rayleigh waves, and Love waves. Rayleigh waves and Love waves are surface waves, where most of the seismic activity is constrained to the surface; P-waves and S-waves extend deeper into the Earth. P-waves ("primary waves") are longitudinal waves, and the fastest of the earthquake waves; they are the waves that are detected first in an earthquake. S-waves ("secondary waves") are transverse waves, vibrating up and down, that travel slower than the P-waves. Rayleigh waves, first predicted by Lord Rayleigh in 1885, are longitudinal surface waves that result in a rolling of the ground, much like the up-and-down motion of waves in the water. Love waves are horizontal transverse waves, shearing the ground side to side as they travel.

With such a wide variety of waves, which can travel with a wide range of wavelengths, it is probably not possible to completely shield a structure from all seismic waves. But a seismic cloak that blocks a significant fraction of seismic waves can complement existing earthquake-resistant structures to make them even more stable during a seismic event.

One additional problem arises with seismic waves: it is not practical, or even moral, to design a cloak that guides earthquake waves around a structure and sends them on their way: the owner of the building directly behind the protected one might be rather unhappy. In 2012, a Korean and an Australian researcher made the first detailed proposal for an earthquake cloak that takes a slightly different approach to protection.[13] The seismic metamaterial, consisting of a large number of metastructures buried in the ground around the building to be protected, would not guide the waves but would instead convert their energy into sound and heat, probably making an incredible racket but saving the building at the center. Consistent with the large wavelengths of earthquake waves, the individual metastructures were envisioned as cylinders that are several meters tall.

But the Korean-Australian research was purely theoretical: Would a seismic cloak work in practice? Also in 2012, German researchers built and tested a prototype of the Marseille-Liverpool acoustic cloak and confirmed its viability. The device was only several centimeters in diameter, but the authors suggested that the design could be scaled up to earthquake-protection sizes.[14]

A more impressive demonstration would soon follow. In 2012, Marseille researchers did the first practical (and successful) test of a seismic metamaterial outside the city of Grenoble, France.[15] A collection of 5-meter-deep holes were drilled to form the metamaterial. Based on simulations, these holes were designed to block seismic waves generated by the "vibrating probe." A collection of sensors was arranged on the opposite side of the metamaterial to see if any waves could penetrate the structure (fig. 48). Their results clearly showed

Figure 48. The seismic metamaterial test done near Grenoble, France, in 2012. Illustration from S. Brule, E. H. Javelaud, S. Enoch, and S. Guenneau, "Experiments on Seismic Metamaterials: Molding Surface Waves," *Physical Review Letters* 112 (2014): 133901. Licensed by CC BY 3.0.

that the metamaterial significantly blocked and attenuated the waves from the source.

But would a seismic metamaterial be effective in an actual earthquake? Here, nature already provided some positive hints. In 2016, researchers from the United Kingdom and France investigated the effectiveness of forests at screening seismic activity. Using ambient seismic noise as a source, the collaborators measured the seismic activity within and just outside a forest in Grenoble. They found that Rayleigh waves were blocked from the forest interior over a wide range of wavelengths. Going further, some of the researchers asked whether it is possible to design a forest to block certain classes of earthquake waves. Using simulations, they determined that an appropriately designed forest could not only obstruct Love waves but even convert

Figure 49. Comparison of the magnetic gradient map of a buried Gallo-Roman theater located at Autun, La Genetoye, France, with the structure of a metamaterial invisibility cloak. Illustration from S. Brule, S. Enoch, and S. Guenneau, "Role of Nanophotonics in the Birth of Seismic Megastructures," *Nanophotonics* 8 (2019): 1591–605. Licensed by CC BY 4.0.

those Love waves into other seismic waves that would propagate harmlessly downward into the Earth. This latter strategy is reminiscent of the sea cloak of Mohammad-Reza Alam, which was envisioned to convert surface ocean waves into deep water waves.[16]

But if appropriately placed trees can suppress earthquake waves, then buildings might serve the same purpose. It had long been noticed that seismic waves travel differently, and can be suppressed, in urban areas. The Marseille research group extended this idea to envision seismic megastructures: entire neighborhoods where buildings are specifically designed and located to act together as a wave-obstructing seismic metamaterial. They were, in effect, imagining an entire invisible city—invisible to seismic waves, at least.

The Marseille researchers had one more surprise in store. While performing research to write their review article on seismic metamaterials, they noticed a striking similarity between the metamaterial designs of invisibility cloaks and the support structures of ancient Gallo-Roman theaters (fig. 49).[17] These theaters were not designed to be seismic cloaks, but their design may have made them inadvertently act as such cloaks. The authors of the review article suggest that this accidental seismic metamaterial structure may have allowed many of these theaters to survive earthquakes that had destroyed other buildings.

But here, in investigating the future of invisibility cloaks in the final chapter of this book, we have come full circle to the beginning. When Perseus was given an invisibility helmet by Zeus in the amphitheater of the city of Joppa in ancient Greece, he may very well have been standing within a seismic invisibility cloak himself!

Will we ever see any of these designs for invisibility and cloaking used in a practical setting? It is clear that many challenges remain to be overcome in order to implement cloaking for any type of waves, and it is possible that they will never be fully overcome. However, as I learned back in 2006 when I incorrectly guessed when the first cloaking experiments would be done, the future of invisibility is very hard to see.

Appendix A: How to Make Your Own Invisibility Device!

The secret of their invisibility lay in their epidermis, corresponding to our skin. This *refracted* all the light between ultraviolet and infrared, the spectrum by which we humans see; carried it clear around them so that to our eyes they appeared perfectly transparent. I have done the same thing with a set of refracting prisms.

Arthur Leo Zagat, "Beyond the Spectrum" (1934)

Note: Some of the experiments listed below contain hazardous materials or materials that could become hazardous if used improperly. Should the reader conduct these experiments, he, she, or they do so at his, her, or their own risk. The author and publisher do not warrant or guarantee the safety of individuals using these procedures and specifically disclaim any and all liability arising directly or indirectly from the use or application of any information contained in this appendix. Should you consider attempting any of these experiments, please make sure you know the proper safety precautions and understand the risks.

As we have seen, true invisibility cloaks require very complicated materials and fabrication processes, and they are likely a long ways from making an appearance, if ever. But there are a number of ways you can have fun with invisibility right in your own home, and it is worth describing a few of them here.

The easiest technique to demonstrate, and one that is truly eye-catching, is refractive index matching. Many craft shops sell bottles of "water beads," which can be used in place of soil for indoor houseplants. The beads are made of a water-absorbing polymer gel and are over 90 percent water when saturated. Their refractive index is therefore extremely close to water, and if you drop them in a glass of water, they will seem to disappear completely. This experiment is reminiscent of the description in *The Invisible Man* of dropping powdered glass into water.

Because the water beads are mostly water anyway, the preceding example may seem like a bit of a cheat for index matching. An alternative experiment can be done with Pyrex glass and mineral oil. Pyrex stirring rods can be ordered inexpensively online, and mineral oil can be purchased at any pharmacy. (Mineral oil is usually used as a laxative, so be prepared for odd looks if you go buying gallons of mineral oil at a time.) The Pyrex has almost the same refractive index as the mineral oil for visible light, and when the Pyrex rod is dipped into the oil, it looks as if it has melted at the surface.

With a little more effort and a different setup, it is even possible to crudely demonstrate the principles of cloaking. (Special thanks to colleagues Mike Fiddy and Robert Ingel for introducing me to this demonstration.) Eight right-angle glass prisms, which can be found online at science supply shops for a few dollars each, are needed to produce the effect. Arrange the prisms as shown (fig. 50). When looked at from the side, one can see objects behind the cloak but cannot see anything placed inside of it.

This cloak takes advantage of the phenomenon of total internal reflection. When light reflects off of the inside surface of glass beyond a critical angle, it is totally reflected. When one looks through the prism cloak in the direction shown in the figure, all the light that is seen is totally reflected at four interior surfaces. Therefore one gets a very clear image of objects behind the cloak but no view of the objects inside.

To make the effect even more pronounced, the prisms can be glued together using an index-matched glass glue, but I've found that simply having the prisms pressed against each other is good enough for demonstrations.

It is also relatively easy at home to explore concepts of transparency, following in the footsteps of Isaac Newton. Paper can be soaked in vegetable or olive oil to make it transparent; as discussed earlier in the book, the oil

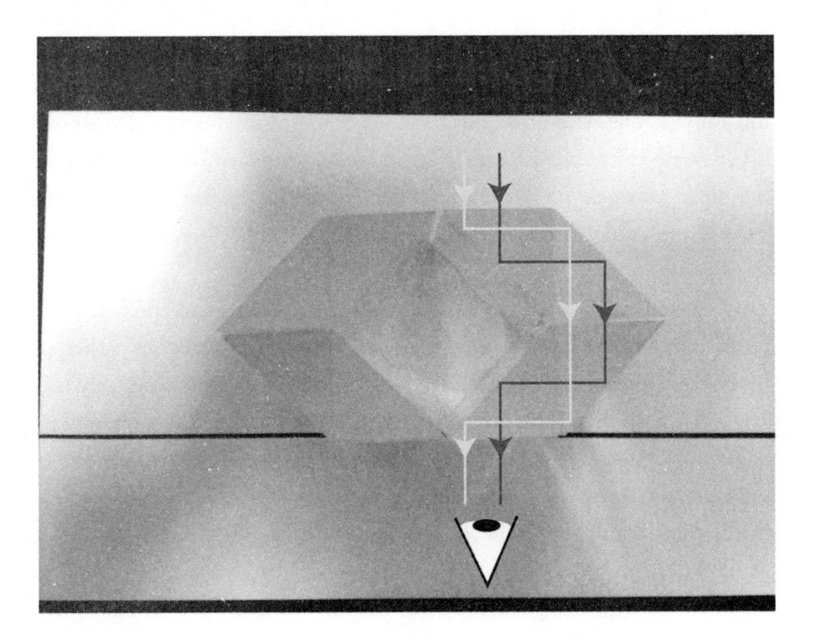

Figure 50. Top view of a prism cloak. When looked at from the direction shown in the illustration, one can see behind the cloak but not within it.

fills the gaps between the paper fibers, reducing the amount of scattering and making it see-through. Another possibility is to purchase Newton's *oculus mundi* (from the Latin "eye of the world"): this stone, also known as lapis mutabilis or hydrophane, is a porous opal that becomes transparent when soaked in water. Again, the water fills the pores of the opal, reducing scattering and increasing transparency. It is possible to purchase samples of hydrophane opal online for about ten dollars.

One other invisibility demonstration is worth mentioning for historical reasons, though I've never successfully done it myself. In 1902, the optical physicist Robert Williams Wood published a paper titled "The Invisibility of Transparent Objects," and it is possibly the first scientific paper to investigate the physics of invisibility, albeit in a limited sense.

This was not a great leap for the imaginative Wood, who was not only an influential optical scientist but also a science fiction author. In 1915, he published, with Arthur Train, *The Man Who Rocked the Earth,* a novel about a rogue scientist who forces all the nations of the world into peace through nuclear weapons. This novel was particularly impressive because it was published some thirty years before nuclear weapons existed. In 1916, the pair published a sequel, *The Moon Maker,* which may be the first science fiction story about sending a spaceship to blow up an asteroid before it collides with the Earth.

In his paper on transparent objects, Wood attempted to confirm a hypothesis by another famous physicist, Lord Rayleigh. Several years earlier, in his paper "Geometrical Objects," Rayleigh had suggested that the only reason that transparent objects can be seen is because they are usually illuminated nonuniformly—that is, more light shines on one side of the object than another. Rayleigh suggested that a transparent object in a region where the light is shining equally in all directions—like an object in a thick fog—would be effectively invisible.

This old hypothesis is almost certainly not true in general, but Wood came up with an ingenious method to test it. To quote from Wood directly,

> I have recently devised a method by which uniform illumination can be very easily obtained and the disappearance of transparent objects when illuminated by it illustrated. The method in brief is to place the object within a hollow globe, the interior surface of which is painted with Balmain's luminous paint and view the interior through a small hole. . . .
>
> If the inner surfaces be exposed to bright daylight, sun or electric light, and the apparatus taken into a dark room, a crystal ball or the cut glass stopper of a decanter placed inside, it will be found to be quite invisible when viewed through the small aperture. A uniform blue glow fills the interior of the ball and only the most careful scrutiny reveals the presence of a solid object within it. One or two of the side facets of the stopper may appear if they happen to reflect or show by refraction any portion of the line of junction of the two hemispheres.

In short: Wood coated the interior of a pair of hemispheres with glow-in-the-dark paint, drilled a small hole through which to observe the interior, and placed a glass object in the joined sphere. According to his results, the glass object could then barely be seen!

I've gone so far as to buy plastic hemispheres and glow-in-the-dark spray paint but have not yet gotten around to assembling the contraption to test it. As we often say in physics textbooks, I leave this as an exercise for the reader!

Appendix B: Invisibibliography

For those who might be interested in reading all the classic invisibility stories, including many that I did not reference in the book proper, I present this "invisibibliography." I have restricted myself largely to science fiction and horror, though some more fantastical stories of note are included. The focus is primarily on stories published before 1960, with a few notable exceptions. Considering how many stories I found through a cursory browsing of old pulp fiction magazines, this list is certainly not complete.

The Invisible Spy, by Explorabilis [Eliza Haywood] (1754). The narrator obtains a belt of invisibility of a dying magus and uses it to get in various misadventures.

The Invisible Gentleman, by James Dalton (1833). A man wishes for invisibility and finds it conferred upon him, leading to all sorts of trouble.

"What Was It?" by Fitz James O'Brien (1859). The first story to present a scientific explanation for invisibility, to explain an invisible monster.

"The Crystal Man," by Edward Page Mitchell (1881). A man gets experimented on and is left in an invisible state, leading to heartbreak.

"The Horla," by Guy de Maupassant (1886). A Parisian finds himself tormented by an invisible being that has taken up residence in his home.

"The Damned Thing," by Ambrose Bierce (1893). A murder inquest leads to the revelation of a being that is of a color unseen.

"Stella," by C. H. Hinton (1895). A peculiar romance, where the executor of an estate finds that one of the homes of the estate is occupied by a woman who was turned invisible so as not to have to deal with the failings of men.

The Invisible Man, by H. G. Wells (1897). The classic. A scientist makes himself invisible and finds it is not as fun as it seemed it would be.

The Secret of William Storitz, by Jules Verne (ca. 1897). A villainous German, spurned in love, takes it upon himself to destroy the future happiness of his would-be bride. Verne's unaltered manuscript was translated into English in 2011 by Bison Books.

"The Shadow and the Flash," by Jack London (1906). Two rival scientists come up with competing methods to turn invisible, and end up meeting a violent end.

"The Thing Invisible," by William Hope Hodgson (1912). One of his famed "Carnacki the Ghost Hunter" tales. Here, Carnacki investigates a chapel supposedly haunted by a murderous ghost. His own experiences convince him that an invisible fiend is attacking people at night. In the end, not a true invisibility story but included because Carnacki spends most of the tale thinking that it is.

The Sea Devils, by Victor Rousseau (1916). A submarine captain learns of the existence of a race of invisible undersea humanoids and struggles to survive their attacks and warn the surface world of their impending invasion. Originally a serial pulp adventure story, it was released as a novel in 1924.

"The People of the Pit," by A. Merritt (1918). A prospector in the Alaskan wilderness comes across a cloaked chasm filled with malevolent inhuman invisible beings.

"The Thing from—'Outside,'" by George Allen England (1923). A party of wilderness explorers quickly realize that they are being stalked by an unseen superintelligent being that views them as little more than objects of research.

"The Monster-God of Mamurth," by Edmond Hamilton (1926). An archaeologist follows ancient inscriptions to a lost city, finding an invisible temple and an immortal invisible monster within.

"The Man Who Could Vanish," by A. Hyatt Verrill (1927). A scientist shows his friend his new method of invisibility, culminating in making a building disappear. Uses optical heterodyning as a mechanism for invisibility, which is a unique explanation.

"Beyond Power of Man," by Paul Ernst (1928). A man agrees to explore an allegedly haunted house and finds himself trapped by the unseen being that lurks there.

"The Dunwich Horror," by H. P. Lovecraft (1929). The corrupt Whateley clan makes a deal with a cosmic horror; when the last member of the clan dies, the invisible being they kept locked up at home goes on a rampage.

"The Shadow of the Beast," by Robert E. Howard (ca. 1930). A man pursues a wanted criminal into a reportedly haunted house. He finds the fugitive dead, and himself hunted by a monster, spirit or beast that cannot be seen, other than its monstrous shadow. Originally unpublished during Howard's lifetime, it was finally printed in 1977.

"The Cave of Horror," by Captain S. P. Meek (1930). When people inexplicably go missing in Mammoth Cave in Kentucky, Dr. Bird is called in to solve the mystery. He finds that an invisible monster from the depths of the Earth has surfaced to feed. A surprisingly fun story with a great description of ultraviolet-based invisibility.

"Invisible Death," by Anthony Pelcher (1930). After an inventor is murdered, a corporation finds itself extorted by someone calling themselves "Invisible Death." They put their top scientist on the case to catch the criminal. This story features the idea of invisibility being generated by vibration, a novel approach!

"The Invisible Master," by Edmond Hamilton (1930). A scientist invents invisibility, gets his device stolen, and soon an invisible criminal is wreaking havoc on the city. The story is not what it appears to be, however, and it contains one of the best descriptions of invisibility and optics I've read in one of these tales.

"The Attack from Space," by Captain S. P. Meek (1930). Beetlelike aliens invade the Earth in order to steal slaves for their radium mines on Mercury. They are nearly unstoppable because they have invisible spacecraft.

"The Invisible Death," by Victor Rousseau (1930). No, you're not seeing double—there were two invisibility stories about an "Invisible Death" in the same year—and the same magazine. The United States is threatened by the "Invisible Emperor" and his armies, which have the power to make themselves, their aircraft, and their buildings invisible. They carve a path of destruction through the country, and it is up to one pilot and a scientist to find a way to stop them.

"Terrors Unseen," by Harl Vincent (1931). Invisible robots. Invisible. Robots. Need I say more? Another story of ultraviolet invisibility.

"The Face in the Abyss," by A. Merritt (1931). An explorer, searching for lost treasure, instead stumbles across a lost civilization, an imprisoned god, and invisible creatures.

"The Murderer Invisible," by Philip Wylie (1931). A mad scientist makes himself invisible and goes on a reign of terror.

"Raiders Invisible," by D. W. Hall (1931). When dirigibles engaged in war-games are mysteriously sabotaged, pilot Chris Travers tracks the cause and finds a conspiracy by the Soviets, armed with the power of invisibility, to destroy the Panama Canal. Another story that uses "Röntgen's rays" to match the refractive index of a body to air.

"The Radiant Shell," by Paul Ernst (1932). When the evil Arvanian government acquires plans to a deadly heat ray, scientist Thorn Winter volunteers to make himself invisible to sneak into the Arvanian embassy to steal the plans.

"The Invisible City," by Clark Ashton Smith (1932). Archaeologists, lost in the desert, stumble across an invisible city and its invisible alien inhabitants. I suspect this one was inspired by Hamilton's, mentioned above.

"Salvage in Space," by Jack Williamson (1933). A space miner comes across a derelict spacecraft and finds it still guarded by a monstrous invisible stowaway.

"Skin and Bones," by Thorne Smith (1933). A man experimenting with fluoroscopic chemicals and a lot of booze makes himself a living skeleton. People are less put off by this than you would think.

"Beyond the Spectrum," by Arthur Leo Zagat (1934). Attempts to drill for water in a growing Florida town instead unleash invisible intelligent monsters from the depths of the Earth.

"The Invisible Bomber," by Lieutenant John Pease (1938). A scientist has created a method to slide between universes, making him invisible, and has outfitted a plane with this technology. He will sell it to the United States . . . for a price.

Sinister Barrier, by Eric Frank Russell (1939). Scientists discover a method to see into the far infrared and find that there are invisible beings controlling humanity. Those beings have no compunctions about killing as many humans as necessary to keep control. A science fiction classic.

"Cloak of Aesir," by Don A. Stuart (1939). In the far future, humanity has been conquered by the alien Sarn. But a mysterious being called Aesir, with a cloak of invulnerable darkness, demands freedom for humankind. The Sarn send agents equipped with invisibility to discover Aesir's secret. One of my favorite invisibility stories.

"In the Walls of Eryx," by H. P. Lovecraft (1939). A prospector on Venus discovers an invisible maze constructed by the natives. His greed gets the better of him and he finds himself trapped within, with his oxygen supply running low.

"The Invisible Robinhood," by Eando Binder (1939). A proto-superhero story about a man who discovers the secret of invisibility in a laboratory accident and uses it to strike fear into the hearts of criminals. The invisibility seems to be explained using the photoelectric effect.

"The Invisible World," by Ed Earl Repp (1940). How about an entire invisible world? Spacefarers suspect that a villainous warlord has a hidden base on an asteroid that is invisible from the outside.

Slan, by A. E. van Vogt (1940). A race of mutants with superior intelligence, strength, and psychic powers, the Slan, struggle to survive in a world where humans exterminate them on sight. The protagonist invents a stealth spaceship that remains cloaked by "disintegrating" any light that touches the hull, a mechanism that sounds strikingly like destructive interference and nonradiating sources.

"The Elixir of Invisibility," by Henry Kuttner (1940). A comedy of invisibility. Richard Rayleigh, the assistant to Dr. Meek, is cajoled into taking Meek's new invisibility elixir as a publicity stunt. But when it turns out that a bank was robbed by an invisible man while Rayleigh was invisible, hijinks ensue.

"Invisible One," by Neil R. Jones (1940). In the twenty-sixth century, a man agrees to be made invisible by cultists in order to rescue his wife, who has been abducted by a space pirate.

"Priestess of the Moon," by Ray Cummings (1940). Beings from the moon—Lunites—use invisibility technology to come steal Earth women. It is as silly as it sounds.

"The Visible Invisible Man," by William P. McGivern (1940). Meek Oscar Doolittle has a mishap with vanishing cream experiments that make him appear and disappear at random. This is especially bad because it happens when he is accused of theft at the bank where he works!

"Land of the Shadow Dragons," by Eando Binder (1941). The return of "The Invisible Robinhood," as he travels to a remote valley filled with invisible animal life—including an invisible T-Rex!

"The Invisible Dove Dancer of Strathpheen Island," by John Collier (1941). An American visiting an island in Ireland becomes convinced that there is an invisible "dove dancer" on the island, visible only by the birds perched on her body, and vows to marry her. Things do not go as planned. Inspired by Rosita Royce, the dove dancer at the 1939 New York World's Fair. As she danced, each dove would fly off with a piece of her clothing in turn.

"Invisible Men of Mars," by Edgar Rice Burroughs (1941). The last "John Carter of Mars" story to be published in Burroughs's lifetime. John Carter and his granddaughter are captured by a city of people that have mastered the science of invisibility!

"The Chameleon Man," by William P. McGivern (1942). A humorous not-quite-invisibility story about a man who is so uninteresting that he literally acquires the chameleonlike ability to blend into his surroundings.

"The Little Man Who Wasn't All There," by Robert Bloch (1942). A comedic story about a man who borrows a magician's topcoat and makes himself partially invisible. The invisibility is vaguely explained as some sort of chemical treatment.

"Ghost Planet," by Thorne Lee (1943). Another invisible planet, but this time the entire planet is literally invisible, not cloaked! The sun of the planet makes everything invisible, and the heroes escape when they get a sunburn that makes them invisible, too.

"The Handyman," by Lester Barclay (1950). A young boy with a strict father makes an invisible friend who can help with the chores and who turns out to be very real. More of a modern fantasy than science fiction.

"Love in the Dark," by H. L. Gold (1951). A woman trapped in an unhappy marriage finds herself being wooed by a new suitor, who happens to be invisible.

"You Can't See Me," by William F. Temple (1951). A man becomes increasingly disturbed as it seems that everyone has made a new invisible best friend except for him. Another "not quite invisibility" story, but a fun science fiction tale.

"War with the Gizmos," by Murray Leinster (1958). Humanity is attacked by a mysterious species of beings that are made of gas, and invisible because of it. A few lucky survivors of early attacks must race to warn civilization.

"The Invisible Man Murder Case," by Henry Slesar (1958). A series of murders are soon suspected to be committed by a man who is not dead, as thought, but invisible.

"For Love," by Aldis Budrys (1962). A titanic alien ship crashes down on Earth for repairs and claims the surface world as its own, leaving humanity hiding underground for decades. In an attempt to gain the upper hand, humankind builds an invisible vehicle to deliver a fusion bomb directly to the otherwise untouchable alien ship.

Memoirs of an Invisible Man, by H. F. Saint (1987). A memoir of a businessman who becomes invisible in a laboratory accident and must figure out how to live his life. Made into a movie directed by John Carpenter in 1992.

Let's Get Invisible!, by R. L. Stine (1993). One of Stine's "Goosebumps" series of books. A young boy finds a mirror that can make him invisible and of course gets into all sorts of trouble.

Mission Invisible, by Ulf Leonhardt (2020). A novel and travel story about science and invisibility by one of the founders of cloaking science.

Acknowledgments

This book project was supposed to be easier than my last one, largely because I am much more familiar with the subject matter. However, the pandemic ended up making everything more difficult, both mentally and emotionally. With that in mind, I would love to thank the many people who helped me get through the past two difficult years, as well as those who helped in various ways to make the book what it is today.

First I thank Beth Szabo, darlene, Damon Diehl, and Taco Visser for being good friends. I also thank, as always, my skating coach, Tappie Dellinger, and my guitar instructor, Toby Watson, for keeping me distracted with fun hobbies. A special thanks to one of my oldest friends, Eric Smith, with apologies for not staying in better touch.

I've been playing a lot of online Dungeons and Dragons during the pandemic, which has kept me from losing my mind, and I thank the four (!) groups I've been playing with; everyone involved is now a good friend, if they weren't already. Thanks to Donna Lanclos, Mindy Weisberger, Dani Marzano, Brad Craddock, Chip Dellinger, and Rachel Parsons of my "Dragon" campaign, and Lali DeRosier, Lisa Manglass, Al Houghton, Josh Witten, Nathan Taylor, and Ashley Gunnet of my "Avernus" campaign. Also special thanks to Hugo González, who runs two games that I play in, and the players of those games: Josh Witten (again), Samantha Hancox-Li, Scott Sutherland, Jean-Sebastien Lodge, and Jim Phoel.

I am part of a vibrant and lovely online community that has been supportive and entertaining in good times and bad. I thank all my online friends for being there for me. I would like to thank a few people in particular for talking with me regularly and being super-supportive: Isla Anderson, Alex Arreola, Nicole Fellouris, Averie Maddox, Lyndell Bade, Brenda Saldana, Samantha Stever, Siiri Takala, Goldie Taylor, Charles Payet, Jacque Gonzales, Brian Malow, Kathy Kerner, Lexie Ali, and @bhaal_spawn. I'm sure I'll remember more people whom I should have included as soon as this book goes to print, so my apologies in advance.

Special thanks to Jim Hathaway, an early and constant supporter of my science communication efforts when I was a beginning professor. Without Jim's support, this book may well have never happened.

I thank Dr. Beth Archer for taking care of my mental and physical health during these past two chaotic years.

As always, I thank my parents, John Gbur and Patricia Gbur, and my sister, Gina Huber, for their love and support.

I reached out to a number of scientists for interviews and information during the writing of the book. I thank Professor John Pendry of Imperial College and Professor Ulf Leonhardt of the Weizmann Institute of Science for generously taking the time to answer questions. I also thank Professor Susumu Tachi of the University of Tokyo and David Smith of Duke University for permission to use selected images.

Special thanks to Wildside Press and the Virginia Kidd Agency for their generous permission to use invisibility quotations from their authors, helping to make the book as I originally envisioned it.

Finally, I thank Mary Pasti, Jean Thomson Black, Laura Jones Dooley, Joyce Ippolito, and Elizabeth Sylvia at Yale University Press for their help in getting this book published and making it the best it can be.

Notes

1. In Which I Made a Bad Prediction

1. Michelson, *Light Waves and Their Uses,* 23–24.

2. "Severe Strain on Credulity."

3. Leonhardt, "Optical Conformal Mapping"; Pendry, Schurig, and Smith, "Controlling Electromagnetic Fields."

4. Schurig et al., "Metamaterial Electromagnetic Cloak."

2. What Do We Mean by "Invisible"?

1. Thone, "Cloaks of Invisibility."

2. "Japanese Scientist Invents 'Invisibility Coat'"; Tachi, "Telexistence and Retroreflective Projection Technology."

3. Brooke, "Tokyo Journal."

4. Diaz, "Teenager Wins $25,000."

5. Mercedes-Benz, "Mercedes-Benz Invisible Car Campaign."

3. Science Meets Fiction

1. Frazer, *Apollodorus* 2.4.2–3.

2. Jowett, *Republic of Plato,* book 2, pp. 37–38.

3. Winter, *Poems and Stories of O'Brien.*

4. Winter.

5. O'Brien, "Lost Room"; O'Brien, "From Hand to Mouth"; O'Brien, "Wondersmith."

6. O'Brien, "Diamond Lens."

7. O'Brien, "What Was It?"

8. Jowett, *Republic of Plato,* book 10, p. 306.

9. It is so common for scientific discoveries to not be named after their original discoverer that the phenomenon has its own tongue-in-cheek law, known as

Stigler's law of eponymy. Appropriately, Stigler himself noted that the law was first introduced, to his knowledge, by the sociologist Robert K. Merton.

10. The law of refraction states $n_1 \sin \theta_1 = n_2 \sin \theta_2$, where n_1 and n_2 are the indices of refraction of the two media and θ_1 and θ_2 are the angles of propagation in the two media; "sin" here represents the trigonometric sine function.

11. London, "Shadow and Flash."

12. Coldewey, "Vantablack."

13. Rogers, "Art Fight!"; Chu, "MIT Engineers Develop 'Blackest Black' Material."

14. Liszewski, "Museum Visitor Falls into Giant Hole."

15. Winter, *Poems and Stories of O'Brien.*

16. Here is one of the weird connections that my mind often makes: the rainbow cover of *Dark Side of the Moon* was created by Storm Thorgerson, a graphic designer who went on in 1983 to direct the music video *Street of Dreams* for the band . . . Rainbow.

17. Newton first introduced the colors as red, orange, yellow, green, blue, indigo, and violet, but today indigo is not recognized as a distinct primary color of the rainbow. You will often see indigo included, however, because it makes for a great mnemonic to remember the order of colors: "Roy G. Biv."

18. Newton, *Opticks,* 249.

4. Invisible Rays, Invisible Monsters

1. Southey, *Doctor.*

2. Smith, *Harmonics;* Smith, *Compleat System of Opticks.*

3. Herschel, "XIII. Investigation of the Powers of the Prismatic Colours to Heat and Illuminate Objects; with Remarks, That Prove the Different Refrangibility of Radiant Heat. To Which Is Added an Inquiry into the Method of Viewing the Sun Advantageously, with Telescopes of Large Apertures and High Magnifying Powers." Herschel was a great scientist but not a concise one.

4. Herschel.

5. Herschel, "XIV. Experiments on Refrangibility."

6. *San Francisco Examiner,* January 21, 1896, 6.

7. Starrett, *Ambrose Bierce,* 22.

8. Bierce, "Damned Thing," 23–24.

9. The first version appeared in *Gil Blas,* October 26, 1886, and the revised version was published by publisher Paul Ollendorff in 1887.

10. Maupassant, *Works of Guy de Maupassant,* 8–9.

11. "Zola's Eulogy," *St. Louis Post Dispatch,* July 30, 1893.

5. Light Comes Out of the Darkness

1. It is fun to talk about the childhood brilliance of people like Young, but it can give a mistaken impression that all scientists must be natural geniuses. In my experience, for every successful scientist who had a gifted childhood there are countless who had unremarkable younger years.

2. Peacock, *Life of Young,* 6.

3. For some context on Bartholomew's age: it was founded only twenty-four years after the First Crusade ended.

4. Young, "Observations on Vision."

5. Young, "Outlines of Experiments and Inquiries respecting Sound and Light."

6. Young, "Mechanism of the Eye."

7. Young, "Outlines of Experiments and Inquiries respecting Sound and Light," 118.

8. Young, "Theory of Light and Colours."

9. Young, 34.

10. Young, "Account of Some Cases of Production of Colours."

11. Young, "Experiments and Calculations relative to Physical Optics."

12. Young, *Course of Lectures on Natural Philosophy.* The use of the terms "Young's double slit experiment" or "Young's two-pinhole experiment" appears to be equally common in the literature, and I wondered for years which way Young himself described it. It turns out, both: "two very small holes or slits."

13. [Brougham],"Bakerian Lecture on Theory of Light and Colours," 450.

14. [Brougham], "Account of Some Cases of Production of Colours," 457.

15. Young, *Reply to the Animadversions of the Edinburgh Reviewers,* 3.

16. Young, 37.

17. Arago, *Biographical Memoir of Young,* 227.

6. Light Goes Sideways

1. Peacock, *Life of Young,* 389. "*Bas bleus*" translates as "blue stockings" and was a term used to describe public intellectual women in the mid-eighteenth century.

2. A classic Slinky toy, which is basically a coiled spring, will also work.

7. Magnets and Currents and Light, Oh My!

1. Young, *Course of Lectures on Natural Philosophy*, 460.

2. You can do this experiment at home with a hot wire Styrofoam cutter, as can be found in hobby stores. The wire carries a direct current and will deflect a compass needle when it is active and brought near.

3. Oersted, "Thermo-Electricity," 717.

4. Oersted, "Experiments on Effect of Current."

5. Jones, *Life and Letters of Faraday*, 1:55.

6. Jones, 1:54.

7. Hirshfeld, *Electric Life of Faraday*, 53.

8. Faraday, "V. Experimental Researches in Electricity."

9. Jones, *Life and Letters of Faraday*, 2:401.

10. Faraday, "III. Experimental Researches in Electricity; Twenty-Eighth Series," 25.

11. Faraday, 26.

12. Campbell and Garnett, *Life of Maxwell*, 28.

13. Anderson, "Forces of Inspiration."

14. Maxwell and Forbes, "On the Description of Oval Curves."

15. Maxwell, "XVIII.—Experiments on Colour."

16. Maxwell, "Faraday's Lines of Force."

17. Maxwell, "XXV. On Physical Lines of Force," 161–62.

18. Maxwell, "III. On Physical Lines of Force," 22.

19. Maxwell, "Dynamical Theory of the Electromagnetic Field."

20. Maxwell, *Treatise on Electricity and Magnetism*, ix.

21. Hertz, *Electric Waves*, 95–106.

22. DeVito, *Science, SETI, and Mathematics*, 49.

8. Waves and Wells

1. Röntgen, "New Kind of Rays."

2. Frankel, "Centennial of Rontgen's Discovery."

3. Ordinary Scotch tape is sticky because it has local regions of positive and negative charge that stick together. In 2008, researchers showed that peeling Scotch tape in a vacuum can produce X-rays, as charges jump from the peeled section of the tape to the smooth side. Without air to slow them down, the electrons produce X-rays when they recombine with the tape.

4. Bostwick, "'Seeing' with X-Rays."

5. Wells, *Experiment in Autobiography,* 53.

6. Wells, 62.

7. Wells, 172.

8. Wells, 254.

9. Wells, 295.

10. Wells, *Invisible Man,* 164.

11. Wells, 171.

12. Wells, *Seven Famous Novels,* viii.

13. Mitchell, "Crystal Man."

14. Hama et al., "Sca*le*."

15. Coxworth, "New Chemical Reagent."

16. Verne, *Secret of Storitz.*

17. Wylie, *Murderer Invisible.*

9. What's in an Atom?

1. Newton, *Opticks,* 394.

2. Nash, "Origin of Dalton's Chemical Atomic Theory."

3. Faraday, "XXIII. Speculation Touching Electric Conduction."

4. The Republic of Ragusa was a small country situated in what is now Croatia that existed from 1358 to 1808, when it was conquered and annexed into the Napoleonic kingdom. No, I didn't know this myself, and I had to look it up.

5. Perrin, "Hypothèses moléculaires," 460.

6. Thomson, "XXIV. Structure of the Atom."

7. Nagaoka, "Kinetics of a System of Particles."

8. Rayleigh, "Electrical Vibrations."

9. Jeans, "Constitution of the Atom."

10. Schott, "Electron Theory of Matter."

11. Einstein, "Über die von der molekularkinetischen Theorie der Wärme geforderte Bewegung von in ruhenden Flüssigkeiten suspendierten Teilchen."

12. Lenard, "Über die Absorption der Kathodenstrahlen verschiedener Geschwindigkeit."

13. Stark, *Prinzipien der Atomdynamik.*

14. Ehrenfest, "Ungleichförmige Elektrizitätsbewegungen ohne Magnet- und Strahlungs-feld."

15. It is somewhat mind-boggling that Rutherford won a Nobel Prize *before* discovering the atomic nucleus.

16. Rutherford, "Forty Years of Physics," 68.

17. Perrin, "Nobel Lecture."

18. Rutherford, "Scattering of α and β Particles by Matter."

10. The Last of the Great Quantum Skeptics

1. Nauenberg, "Max Planck," 715.

2. Stephen Hawking was once told that every equation that he included in a popular science book would halve its sales. With that in mind, my apologies.

3. Nauenberg, "Max Planck," 715.

4. Wheaton, "Philipp Lenard."

5. Einstein, "Über einen die Erzeugung und Verwandlung des Lichtes betreffenden heuristischen Gesichtspunkt."

6. Don't worry if the idea of wave-particle duality doesn't make any sense to you; in fact, physicists to this day are still struggling with understanding what, exactly, wave-particle duality means.

7. Niaz et al., "History of the Photoelectric Effect," 909.

8. Bohr, "Constitution of Atoms and Molecules."

9. Broglie, "Wave Nature of the Electron."

10. My late PhD adviser Emil Wolf often jokingly said of one of his students that "he causes more problems than he solves." This was not meant as an insult, however: the student was so insightful that he regularly found numerous new questions to investigate in the course of solving a single problem. This is the ideal for science: answer questions and in the process find new questions to answer.

11. Students first learning quantum physics are often taught the mantra "Shut up and calculate." In other words, they're told not to worry about *why* the physics works but just to use it without question. More than a century after the era of quantum physics began, we still don't know exactly what it tells us about the nature of the universe. To borrow a phrase from Douglas Adams and Mark Carwardine's book *Last Chance to See,* "It was hard to avoid the feeling that somebody, somewhere, was missing the point. I couldn't even be sure that it wasn't me" (153).

12. Schott, "V. Reflection and Refraction of Light."

13. Schott, "LIX. Radiation from Moving Systems of Electrons," 667.

14. Schott, "XXII. Bohr's Hypothesis of Stationary States of Motion," 258, 243.

15. Schott, "LIX. Electromagnetic Field of a Moving Uniformly and Rigidly Electrified Sphere."

16. Schott, 752–53.

17. Schott, "Electromagnetic Field due to a Uniformly and Rigidly Electrified Sphere in Spinless Accelerated Motion and Its Mechanical Reaction on the Sphere," I, II, III, and IV.

18. Conway, "Professor G. A. Schott, 1868–1937."

19. Bohm and Weinstein, "Self-Oscillations of a Charged Particle."

20. Goedecke, "Classically Radiationless Motions," B288.

11. Seeing Inside

1. Bostwick "'Seeing' with X-Rays."

2. "Edison Says There Is Hope."

3. "Edison Fears Hidden Perils of the X-Rays."

4. Smith, *Skin and Bones.* Thorne Smith is most famous for the 1926 novel *Topper,* about a banker and his wife becoming friends with a couple of ghosts. The novel was made into a movie in 1937.

5. Gernsback, "Can We Make Ourselves Invisible?"

6. Cormack, "Nobel Lecture."

7. Cormack.

8. If EMI sounds familiar, it is the same company that produced some of the most significant musical albums of the 1960s and 1970s, including work by the Beatles and Pink Floyd.

9. Hounsfield, "Computerized Transverse Axial Scanning (Tomography)."

10. Weyl, "Über die asymptotische Verteilung der Eigenwerte."

11. Wikipedia, s.v. "Invers Problem."

12. Bleistein and Bojarski, "Recently Developed Formulations of the Inverse Problem," 1–2.

13. Moses, "Solution of Maxwell's Equations," 1670.

14. Bleistein and Cohen, "Nonuniqueness in the Inverse Source Problem."

15. Bojarski, "Inverse Scattering Inverse Source Theory."

16. Stone, "Nonradiating Sources of Compact Support Do Not Exist."

17. Devaney and Sherman, "Nonuniqueness in Inverse Source and Scattering Problems."

18. Devaney and Sherman, 1041–42.

12. A Wolf on the Hunt

Note to epigraph: F. Scarlett Potter, "The Were-Wolf of the Grendelwold" (1882), reprinted in Easley and Scott, *Terrifying Transformations.*

1. Kerker, "Invisible Bodies."

2. From recollections of personal conversations with Emil Wolf.

3. From recollections of personal conversations with Emil Wolf.

4. Wolf later recalled that Born was also avoiding copyright issues with his old book. As a German Jew, Born had the rights of his book seized by the Nazis when they came to power, and when the Nazis were defeated, the Allies claimed the rights to his book as spoils of war. When Born announced he was working on a new optics book, the rights-holder contacted him to ask which parts of his old book he was using, so they could bill him for their use; he essentially told them to stuff it.

5. From recollections of personal conversations with Emil Wolf.

6. Wolf, "Optics in Terms of Observable Quantities."

7. Wolf, "Recollections of Max Born," 12.

8. Wolf, 15.

9. Wolf, 15.

10. Wolf, "Three-Dimensional Structure Determination of Semi-Transparent Objects."

11. One of my great regrets in working with Emil Wolf is that I never asked him exactly how he became interested in nonradiating sources.

12. Devaney and Wolf, "Radiating and Nonradiating Classical Current Distributions." Devaney would go on to be a pioneer in the use of diffraction tomography for seismic exploration.

13. Kim and Wolf, "Non-Radiating Monochromatic Sources"; Gamliel et al., "New Method for Specifying Nonradiating Monochromatic Sources."

14. Gbur, "Sources of Arbitrary States of Coherence."

15. Gbur, "Nonradiating Sources and the Inverse Source Problem."

16. Devaney, "Nonuniqueness in the Inverse Scattering Problem."

17. Wolf and Habashy, "Invisible Bodies."

18. Nachman, "Reconstructions from Boundary Measurements."

19. Wolf, "Recollections of Max Born," 15.

13. Materials Not Found in Nature

1. Wiener, "Stehende Lichtwellen," 240–41.

2. Sommerfeld, *Optics,* 18.

3. Pendry et al., "Extremely Low Frequency Plasmons."

4. Pendry et al., "Magnetism from Conductors."

5. Smith et al., "Composite Medium"; Smith and Kroll, "Negative Refractive Index"; Shelby et al., "Experimental Verification."

6. Vesalago, "Electrodynamics of Substances." Fortunately, Victor Vesalago lived to see his work widely celebrated. Before his death in 2018, he was invited to numerous international conferences to give talks on his ideas. At one meeting, I happened to see him sitting alone at one point and took the opportunity to introduce myself and thank him for his contributions.

7. Pendry, "Negative Refraction Makes a Perfect Lens."

8. Grbic and Eleftheriades, "Overcoming the Diffraction Limit."

9. Fang et al., "Sub-Diffraction-Limited Optical Imaging."

14. Invisibility Cloaks Appear

1. Pendry and Ramakrishna, "Near-Field Lenses in Two Dimensions"; Pendry, "Perfect Cylindrical Lenses."

2. Ward and Pendry, "Refraction and Geometry in Maxwell's Equations."

3. Ball, "Bending the Laws of Optics with Metamaterials," 201.

4. Leonhardt and Piwnicki, "Optics of Nonuniformly Moving Media."

5. Leonhardt, private correspondence.

6. Leonhardt, private correspondence.

7. Petit, "Invisibility Uncloaked."

8. Leonhardt, "Optical Conformal Mapping"; Pendry et al., "Controlling Electromagnetic Fields."

9. Ball, "Invisibility Cloaks Are in Sight."

10. Merritt, *Face in the Abyss.*

11. Ulf Leonhardt has been extremely active in invisibility research and transformation optics since 2006, including writing a novel about invisibility, *Mission Invisible* (2020).

12. Schurig et al., "Metamaterial Electromagnetic Cloak at Microwave Frequencies."

13. Boyle, "Here's How to Make an Invisibility Cloak."

15. Things Get Weird

1. Hashemi et al., "Diameter-Bandwidth Product Limitation of Isolated-Object Cloaking."

2. Li and Pendry, "Hiding under the Carpet."

3. Li et al., "Broadband Ground-Plane Cloak."

4. Valentine et al., "Optical Cloak Made of Dielectrics."

5. Ergin et al., "Three-Dimensional Invisibility Cloak at Optical Wavelengths."

6. Zhang et al., "Macroscopic Invisibility Cloak for Visible Light."

7. Chen et al., "Macroscopic Invisibility Cloaking of Visible Light."

8. Sinclair, "Invisibility Cloak Demoed at TED2013."

9. Chen et al., "Ray-Optics Cloaking Devices for Large Objects."

10. Ball, "'Invisibility Cloak' Hides Cats and Fish."

11. Alù and Engheta, "Achieving Transparency with Plasmonic and Metamaterial Coatings."

12. Alù and Engheta, "Cloaking a Sensor."

13. Lai et al., "Complementary Media Invisibility Cloak."

14. Lai et al., "Illusion Optics."

15. Luo et al., "Conceal an Entrance by Means of Superscatterer."

16. Greenleaf et al., "Electromagnetic Wormholes and Virtual Magnetic Monopoles."

17. Greenleaf et al., "Anisotropic Conductivities That Cannot Be Detected by EIT."

18. Prat-Camps, Navau, and Sanchez, "Magnetic Wormhole."

19. This is a standard demonstration in physics, and one can buy bar magnets from science supply stores that are easily breakable to show that each piece ends up with its own North and South poles.

20. Dirac, "Quantised Singularities in the Electromagnetic Field."

21. Milton, "Theoretical and Experimental Status of Magnetic Monopoles."

22. Chen et al., "Anti-Cloak."

23. Tsakmakidis et al., "Ultrabroadband 3D Invisibility with Fast-Light Cloaks."

16. More Than Hiding

1. In fact, optics researchers have been using light waves as a substitute for water waves as a way to study reliably and safely how rogue waves occur.

2. Dysthe, Krogstad, and Müller, "Oceanic Rogue Waves."

3. Alam, "Broadband Cloaking in Stratified Seas."

4. At this point I should note that there are many, many results that have come out about cloaking of various types since 2006. The best we can do here is highlight some interesting results. My apologies to those authors whose work is not mentioned—it is not a judgment on the research at all.

5. Kwon and Werner, "Transformation Optical Designs for Wave Collimators."

6. Rahm et al., "Optical Design of Reflectionless Complex Media"; Markov, Valentine, and Weiss, "Fiber-to-Chip Coupler."

7. Wood and Pendry, "Metamaterials at Zero Frequency"; Magnus et al., "A d.c. Magnetic Material."

8. Navau et al., "Magnetic Properties of a dc Metamaterial"; Gömöry et al., "Experimental Realization of a Magnetic Cloak."

9. Yang et al., "dc Electric Invisibility Cloak."

10. Guenneau, Amra, and Veyante, "Transformation Thermodynamics."

11. Cummer et al., "Scattering Theory Derivation of a 3D Acoustic Cloaking Shell"; Zhang, Xia, and Fang, "Broadband Acoustic Cloak for Ultrasound Waves."

12. Farhat, Guenneau, and Enoch, "Ultrabroadband Elastic Cloaking in Thin Plates."

13. Kim and Das, "Seismic Waveguide of Metamaterials."

14. Stenger, Wilhelm, and Wegener, "Experiments on Elastic Cloaking in Thin Plates."

15. Brûlé et al., "Experiments on Seismic Metamaterials."

16. Colombi et al., "Forests as a Natural Seismic Metamaterial"; Maruel et al., "Conversion of Love Waves in a Forest of Trees."

17. Brûlé, Enoch, and Guenneau, "Role of Nanophotonics in the Birth of Seismic Megastructures."

Bibliography

Adams, Douglas, and Mark Carwardine. *Last Chance to See.* New York: Ballantine Books, 1992.

Alam, Mohammad-Reza. "Broadband Cloaking in Stratified Seas." *Physical Review Letters* 108 (2012): 084502.

Alù, Andrea, and Nader Engheta. "Achieving Transparency with Plasmonic and Metamaterial Coatings." *Physical Review E* 72 (2005): 016623.

———. "Cloaking a Sensor." *Physical Review Letters* 102 (2009): 233901.

Anderson, Anthony F. "Forces of Inspiration." *New Scientist,* June 11, 1981, 712–13.

Apollodorus. *The Library.* Trans. J. G. Frazer. London: William Heinemann, 1921.

Arago, M. "Biographical Memoir of Dr. Thomas Young." *Edinburgh New Philosophical Journal* 20 (1836): 213–40.

Ball, Philip. "Bending the Laws of Optics with Metamaterials: An Interview with John Pendry." *National Science Review* 5 (2018): 200–202.

———. "'Invisibility Cloak' Hides Cats and Fish." *Nature,* June 11, 2013. https://doi.org/10.1038/nature.2013.13184.

———. "Invisibility Cloaks Are in Sight." *Nature News,* May 25, 2006. https://doi.org/10.1038/news060522-18.

Bierce, Ambrose. "The Damned Thing." *Town Topics* (New York), December 7, 1893.

Bleistein, Norman, and Norbert N. Bojarski. "Recently Developed Formulations of the Inverse Problem in Acoustics and Electromagnetics." Denver Research Institute, Division of Mathematical Sciences, 1974.

Bleistein, Norman, and Jack K. Cohen. "Nonuniqueness in the Inverse

Source Problem in Acoustics and Electromagnetics." *Journal of Mathematical Physics* 18 (1977): 194–201.

Bohm, D., and M. Weinstein. "The Self-Oscillations of a Charged Particle." *Physical Review* 74 (1948): 1789–98.

Bohr, Niels. "On the Constitution of Atoms and Molecules." *Philosophical Magazine* 26 (1913): 1–24.

Bojarski, Norbert N. "Inverse Scattering Inverse Source Theory." *Journal of Mathematical Physics,* 22 (1981): 1647–50.

Bostwick, A. E. "'Seeing' with X-Rays." *Courier-News* (Bridgewater, N.J.), May 27, 1896, 7.

Boyle, Alan. "Here's How to Make an Invisibility Cloak." *NBC News,* May 25, 2006, www.nbcnews.com/id/wbna12961080.

Broglie, Louis de. "The Wave Nature of the Electron." In *Nobel Lectures: Physics, 1922–1941,* 244–56. Amsterdam: Elsevier, 1965.

Brooke, James. "Tokyo Journal; Behold, the Invisible Man, If Not Seeing Is Believing." *New York Times,* March 27, 2003.

[Brougham, Henry]. "An Account of Some Cases of the Production of Colours Not Hitherto Described." *Edinburgh Review* 1 (1803): 457–60.

———. "The Bakerian Lecture on the Theory of Light and Colours." *Edinburgh Review* 1 (1803): 450–56.

Brûlé, Stéphane, Stefan Enoch, and Sébastien Guenneau. "Role of Nanophotonics in the Birth of Seismic Megastructures." *Nanophotonics* 8 (2019): 1591–605.

Brûlé, S., E. H. Javelaud, S. Enoch, and S. Guenneau. "Experiments on Seismic Metamaterials: Molding Surface Waves." *Physical Review Letters* 112 (2014): 133901.

Campbell, Lewis, and William Garnett. *The Life of James Clerk Maxwell.* London: Macmillan, 1882.

Castaldi, Giuseppe, Ilaria Gallina, Vincenzo Galdi, Andrea Alù, and Nader Engheta. "Cloak/Anti-Cloak Interactions." *Optics Express* 17 (2009): 3101–14.

Chen, Huanyang, and C. T. Chan. "Acoustic Cloaking in Three Dimensions Using Acoustic Metamaterials." *Applied Physics Letters* 91 (2007): 183518.

Chen, Huanyang, Xudong Luo, Hongru Ma, and C. T. Chan. "The Anti-Cloak." *Optics Express* 16 (2008): 14603–8.

Chen, Huanyang, Rong-Xin Miao, and Miao Li. "Transformation Optics That Mimics the System outside a Schwarzschild Black Hole." *Optics Express* 18 (2010): 15183–88.

Chen, Huanyang, Bae-Ian Wu, Baile Zhang, and Jin Au Kong. "Electromagnetic Wave Interactions with a Metamaterial Cloak." *Physical Review Letters* 99 (2007): 063903.

Chen, Huanyang, Bin Zheng, Lian Shen, Huaping Wang, Xianmin Zhang, Nikolay I. Zheludev, and Baile Zhang. "Ray-Optics Cloaking Devices for Large Objects in Incoherent Natural Light." *Nature Communications* 4 (2013): 2652.

Chen, Xianzhong, Yu Luo, Jingjing Zhang, Kyle Jiang, John B. Pendry, and Shuang Zhang. "Macroscopic Invisibility Cloaking of Visible Light." *Nature Communications* 2 (2011): 176.

Cho, Adrian. "High-Tech Materials Could Render Objects Invisible." *Science* 312 (2006): 1120.

Chu, Jennifer. "MIT Engineers Develop 'Blackest Black' Material to Date." *MIT News*, September 12, 2019. http://news.mit.edu/2019/blackest-black-material-cnt-0913.

Coldewey, Devin. "Vantablack: U.K. Firm Shows Off 'World's Darkest Material.'" *NBC News*, July 14, 2014. www.nbcnews.com/science/science-news/vantablack-u-k-firm-shows-worlds-darkest-material-n155581.

Colombi, Andrea, Philippe Roux, Sébastien Guenneau, Philippe Gueguen, and Richard V. Craster. "Forests as a Natural Seismic Metamaterial: Rayleigh Wave Bandgaps Induced by Local Resonances." *Scientific Reports* 6 (2016): 19238.

Conway, Arthur William. "Professor G. A. Schott, 1868–1937." *Obituary Notices of Fellows of the Royal Society* 2 (1939): 451–54.

Cormack, Allan M. "Nobel Lecture." *The Nobel Prize*, www.nobelprize.org/prizes/medicine/1979/cormack/lecture/.

Coxworth, Ben. "New Chemical Reagent Turns Biological Tissue Transparent." *New Atlas*, September 2, 2011. https://newatlas.com/chemical-reagent-turns-biological-tissue-transparent/19708/.

Cummer, Steven A., Bogdan-Ioan Popa, David Schurig, David R. Smith, John Pendry, Marco Rahm, and Anthony Starr. "Scattering Theory Derivation of a 3D Acoustic Cloaking Shell." *Physical Review Letters* 100 (2008): 024301.

Cummer, Steven A., and David Schurig. "One Path to Acoustic Cloaking." *New Journal of Physics* 9 (2007): 45.

Devaney, A. J. "Nonuniqueness in the Inverse Scattering Problem." *Journal of Mathematical Physics* 19 (1978): 1526–31.

Devaney, A. J., and G. C. Sherman. "Nonuniqueness in Inverse Source and Scattering Problems." *IEEE Transactions on Antennas and Propagation* 30 (1982): 1034–37.

Devaney, A. J., and E. Wolf. "Radiating and Nonradiating Classical Current Distributions and the Fields They Generate." *Physical Review D* 8 (1973): 1044–47.

DeVito, Carl L. *Science, SETI, and Mathematics.* New York: Berghahn Books, 2014.

Diaz, Johnny. "Teenager Wins $25,000 for Science Project That Solves Blind Spots in Cars." *New York Times,* November 7, 2019.

Dirac, Paul Adrien Maurice. "Quantised Singularities in the Electromagnetic Field." *Proceedings of the Royal Society A* 133 (1931): 60–72.

Doyle, A. Conan. "The Adventure of the Abbey Grange." *Strand,* September 1904, 243–56.

Dysthe, Kristian, Harald E. Krogstad, and Peter Müller. "Oceanic Rogue Waves." *Annual Review of Fluid Mechanics* 40 (2008): 287–310.

Easley, Alexis, and Shannon Scott, eds. *Terrifying Transformations: An Anthology of Victorian Werewolf Fiction.* Kansas City, Mo.: Valancourt Books, 2013.

"Edison Fears Hidden Perils of the X-Rays." *New York World,* August 3, 1903.

"Edison Says There Is Hope." *San Francisco Examiner,* November 19, 1896, 5.

Ehrenfest, Paul. "Ungleichförmige Elektrizitätsbewegungen ohne Magnet- und Strahlungsfeld." *Physikalische Zeitschrift* 11 (1910): 708–9.

Einstein, A."Über die von der molekularkinetischen Theorie der Wärme geforderte Bewegung von in ruhenden Flüssigkeiten suspendierten Teilchen." *Annalen der Physik* 332 (1905): 549–60.

———. "Über einen die Erzeugung und Verwandlung des Lichtes betreffenden heuristischen Gesichtspunkt." *Annalen der Physik* 332 (1905): 132–48.

Ergin, Tolga, Nicholas Stenger, Patrice Brenner, John B. Pendry, and Martin Wegener. "Three-Dimensional Invisibility Cloak at Optical Wavelengths." *Science* 328 (2010): 337–39.

Fang, Nicholas, Hyesog Lee, Cheng Sun, and Xiang Zhang. "Sub-Diffraction-Limited Optical Imaging with a Silver Superlens." *Science* 308 (2005): 534–37.

Faraday, Michael. "III. Experimental Researches in Electricity, Twenty-Eighth Series." *Philosophical Transactions of the Royal Society of London* 142 (1852): 25–56.

———. "V. Experimental Researches in Electricity." *Philosophical Transactions of the Royal Society of London* 122 (1832): 125–62.

———. "XXIII. A Speculation Touching Electric Conduction and the Nature of Matter." *London, Edinburgh, and Dublin Philosophical Magazine and Journal of Science* 24 (1844): 136–44.

Farhat, M., S. Enoch, S. Guenneau, and A. B. Movchan. "Broadband Cylindrical Acoustic Cloak for Linear Surface Waves in a Fluid." *Physical Review Letters* 101 (2008): 134501.

Farhat, M., S. Guenneau, and S. Enoch. "Ultrabroadband Elastic Cloaking in Thin Plates." *Physical Review Letters* 103 (2009): 024301.

Frankel, R. I. "Centennial of Rontgen's Discovery of X-Rays." *Western Journal of Medicine* 164 (1996): 497–501.

Gamliel, A., K. Kim, A. I. Nachman, and E. Wolf. "A New Method for Specifying Nonradiating Monochromatic Sources and Their Fields." *Journal of the Optical Society of America A* 6 (1989): 1388–93.

García-Meca, C., M. M. Tung, J. V. Galán, R. Ortuño, F. J. Rodríguez-Fortuno, J. Martí, and A. Martínez. "Squeezing and Expanding Light without Reflections via Transformation Optics." *Optics Express* 19 (2011): 3562–75.

Gbur, Greg. "Nonradiating Sources and the Inverse Source Problem." Ph.D. thesis, University of Rochester, 2001.

Gbur, Greg, and Emil Wolf. "Sources of Arbitrary States of Coherence That

Generate Completely Coherent Fields outside the Source." *Optics Letters* 22 (1997): 943–45.

Genov, Dentcho A., Shuang Zhang, and Xiang Zhang. "Mimicking Celestial Mechanics in Metamaterials." *Nature Physics* 5 (2009): 687–92.

Gernsback, H. "Can We Make Ourselves Invisible?" *Science and Invention* 8 (1921): 1074.

Goedecke, G. H. "Classically Radiationless Motions and Possible Implications for Quantum Theory." *Physical Review* 135 (1964): B281–88.

Gömöry, Fedor, Mykola Solovyov, Ján Šouc, Carles Navau, Jordi Prat-Camps, and Alvaro Sanchez. "Experimental Realization of a Magnetic Cloak." *Science* 335 (2012): 1466–68.

Gonzalez, Robbie. "A Chemical That Can Turn Your Organs Transparent." *Gizmodo,* September 1, 2011. https://gizmodo.com/a-chemical-that-can-turn-your-organs-transparent-5836605.

Grbic, Anthony, and George V. Eleftheriades. "Overcoming the Diffraction Limit with a Planar Left-Handed Transmission-Line Lens." *Physical Review Letters* 92 (2004): 117403.

Greenleaf, Allan, Yaroslav Kurylev, Matti Lassas, and Gunther Uhlmann. "Electromagnetic Wormholes and Virtual Magnetic Monopoles from Metamaterials." *Physical Review Letters* 99 (2007): 183901.

Greenleaf, Allan, Matti Lassas, and Gunther Uhlmann. "Anisotropic Conductivities That Cannot Be Detected by EIT." *Physiological Measurement* 24 (2003): 413–19.

Guenneau, Sebastien, Claude Amra, and Denis Veynante. "Transformation Thermodynamics: Cloaking and Concentrating Heat Flux." *Optics Express* 20 (2012): 8207–18.

Hama, Hiroshi, Hiroshi Kurokawa, Hioyuki Kawano, Ryoko Ando, Tomomi Shimogori, Hisayori Noda, Kiyoko Fukami, Asako Sakaue-Sawano, and Atsushi Miyawaki. "Sca*l*e: A Chemical Approach for Fluorescence Imaging and Reconstruction of Transparent Mouse Brain." *Nature Neuroscience* 14 (2011): 1481–88.

Hapgood, Fred, and Andrew Grant. "Metamaterial Revolution: The New Science of Making Anything Disappear." *Discover Magazine,* April 2009.

Hashemi, Hila, Cheng Wei Qiu, Alexander P. McCauley, J. D. Joannopoulos, and Steven G. Johnson. "Diameter-Bandwidth Product Limitation of Isolated-Object Cloaking." *Physical Review A* 86 (2012): 013804.

Herschel, William. "XIII. Investigation of the Powers of the Prismatic Colours to Heat and Illuminate Objects; with Remarks, That Prove the Different Refrangibility of Radiant Heat. To Which Is Added an Inquiry into the Method of Viewing the Sun Advantageously, with Telescopes of Large Apertures and High Magnifying Powers." *Philosophical Transactions of the Royal Society of London* 90 (1800): 255–83.

———. "XIV. Experiments on the Refrangibility of the Invisible Rays of the Sun." *Philosophical Transactions of the Royal Society of London* 90 (1800): 284–92.

Hertz, Heinrich. *Electric Waves; Being Researches on the Propagation of Electric Action with Finite Velocity through Space.* Translated by D. E. Jones. 1895. Reprint, New York: Dover, 1962.

Hirshfeld, Alan. *The Electric Life of Michael Faraday.* New York: Walker, 2006.

Hounsfield, G. N. "Computerized Transverse Axial Scanning (Tomography): Part I. Description of System." *British Journal of Radiology* 46 (1973): 1016–22.

James, R. W. *The Optical Principles of the Diffraction of X-Rays.* London: G. Bell and Sons, 1948.

"Japanese Scientist Invents 'Invisibility Coat.'" *BBC News World Edition,* February 18, 2003. http://news.bbc.co.uk/2/hi/asia-pacific/2777111.stm.

Jeans, J. H. "On the Constitution of the Atom." *Philosophical Magazine* 11 (1906): 604–7.

Jiang, Wei Xiang, Hui Feng Ma, Qiang Cheng, and Tie Jun Cui. "Illusion Media: Generating Virtual Objects Using Realizable Metamaterials." *Applied Physics Letters* 96 (2010): 121910.

Jones, Bence. *The Life and Letters of Faraday.* 2 vols. Philadelphia: J. B. Lippincott, 1870.

Jowett, Benjamin, trans. *The Republic of Plato.* 2nd ed. Oxford: Clarendon Press, 1881.

Kaye, G. W. C. *X Rays.* 3rd ed. London: Longmans, Green, 1918.

Kerker, Milton. "Invisible Bodies." *Journal of the Optical Society of America* 65 (1975): 376–79.

Kim, Kisik, and Emil Wolf. "Non-Radiating Monochromatic Sources and Their Fields." *Optics Communications* 59 (1986): 1–6.

Kim, Sang-Hoon, and Mukunda P. Das. "Seismic Waveguide of Metamaterials." *Modern Physics Letters B* 26 (2012): 1250105.

Kwon, Do-Hoon, and Douglas H. Werner. "Transformation Optical Designs for Wave Collimators, Flat Lenses and Right-Angle Bends." *New Journal of Physics* 10 (2008): 115023.

Lai, Yun, Huanyang Chen, Zhao-Qing Zhang, and C. T. Chan. "Complementary Media Invisibility Cloak That Cloaks Objects at a Distance Outside the Cloaking Shell." *Physical Review Letters* 102 (2009): 093901.

Lai, Yun, Jack Ng, HuanYang Chen, DeZhuan Han, JunJun Xiao, Zhao-Qing Zhang, and C. T. Chan. "Illusion Optics: The Optical Transformation of an Object into Another Object." *Physical Review Letters* 102 (2009): 253902.

Lenard, P. "Über die Absorption der Kathodenstrahlen verschiedener Geschwindigkeit." *Annalen der Physik* 12 (1903): 714–44.

Leonhardt, Ulf. "Optical Conformal Mapping." *Science* 312 (2006): 1777–80.

Leonhardt, U., and P. Piwnicki. "Optics of Nonuniformly Moving Media." *Physical Review A* 60 (1999): 4301–12.

Li, Jensen, and J. B. Pendry. "Hiding under the Carpet: A New Strategy for Cloaking." *Physical Review Letters* 101 (2008): 203901.

Lie, R., C. Ji, J. J. Mock, J. Y. Chin, T. J. Cui, and D. R. Smith. "Broadband Ground-Plane Cloak." *Science* 323 (2009): 366–69.

Liszewski, Andrew. "Museum Visitor Falls into Giant Hole That Looks Like a Cartoonish Painting on the Floor." *Gizmodo,* August 20, 2018. https://gizmodo.com/museum-visitor-falls-into-giant-hole-that-looks-like-a-1828462859.

London, Jack. "The Shadow and the Flash." *Windsor Magazine* 24 (1906): 354–62.

Luo, Xudong, Tao Yang, Yongwei Gu, Huanyang Chen, and Hongru Ma. "Conceal an Entrance by Means of Superscatterer." *Applied Physics Letters* 94 (2009): 223513.

Magnus, F., B. Wood, J. Moore, K. Morrison, G. Perkins, J. Fyson, M. C. K.

Wiltshire, D. Caplin, L. F. Cohen, and J. B. Pendry. "A d.c. Magnetic Material." *Nature Materials* 7 (2008): 295–97.

Markov, Petr, Jason G. Valentine, and Sharon M. Weiss. "Fiber-to-Chip Coupler Designed Using an Optical Transformation." *Optics Express* 20 (2012): 14705–13.

Maupassant, Guy de. "Le Horla." *Gil Blas,* October 26, 1886.

———. *Le Horla.* Paris: Paul Ollendorff, 1887.

———. *The Works of Guy de Maupassant.* Vol. 4. London: Classic, 1911.

Maurel, Agnes, Jean-Jacques Marigo, Kim Pham, and Sebastien Guenneau. "Conversion of Love Waves in a Forest of Trees." *Physical Review B* 98 (2018): 134311.

Maxwell, J. C. "A Dynamical Theory of the Electromagnetic Field." *Philosophical Transactions of the Royal Society of London* 155 (1865): 459–512.

———. "On Faraday's Lines of Force." *Transactions of the Cambridge Philosophical Society* 10 (1855): 155–229.

———. *A Treatise on Electricity and Magnetism.* 3rd ed. Oxford: Clarendon Press, 1892.

———. "III. On Physical Lines of Force." *London, Edinburgh, and Dublin Philosophical Magazine and Journal of Science* 23 (1862): 12–24.

———. "XVIII.—Experiments on Colour, as Perceived by the Eye, with Remarks on Colourblindness." *Transactions of the Royal Society of Edinburgh* 21 (1857): 275–98.

———. "XXV. On Physical Lines of Force." *London, Edinburgh, and Dublin Philosophical Magazine and Journal of Science* 21 (1861): 161–75.

Maxwell, J. C., and Forbes. "1. On the Description of Oval Curves, and Those Having a Plurality of Foci." *Proceedings of the Royal Society of Edinburgh* 2 (1951): 89–91.

Mercedes-Benz. "Mercedes-Benz Invisible Car." 2012. https://www.youtube.com/watch?v=TYlXpnPTbqQ.

Merritt, A. *The Face in the Abyss.* New York: Horace Liveright, 1931.

Michelson, A. A. *Light Waves and Their Uses.* Chicago: University of Chicago Press, 1903.

Milton, K. A. "Theoretical and Experimental Status of Magnetic Monopoles." *Reports on Progress in Physics* 69 (2006): 1637–711.

Mitchell, Edward Page. "The Crystal Man." *New York Sun*, January 30, 1881.

Mizuno, Kohei, Juntaro Ishii, Hideo Kishida, Yuhei Hayamizu, Satoshi Yasuda, Don N. Futaba, Motoo Yumura, and Kenji Hata. "A Black Body Absorber from Vertically Aligned Single-Walled Carbon Nanotubes." *Proceedings of the National Academy of Sciences* 106 (2009): 6044–47.

Monticone, Francesco, and Andrea Alù. "Do Cloaked Objects Really Scatter Less?" *Physical Review X* 3 (2013): 041005.

Moses, H. E. "Solution of Maxwell's Equations in Terms of a Spinor Notation: The Direct and Inverse Problem." *Physical Review* 113 (1959): 1670–79.

Nachman, Adrian I. "Reconstructions from Boundary Measurements." *Annals of Mathematics* 128 (1988): 531–76.

Nagaoka, H. "LV. Kinetics of a System of Particles Illustrating the Line and the Band Spectrum and the Phenomena of Radioactivity." *London, Edinburgh, and Dublin Philosophical Magazine and Journal of Science* 7 (1904): 445–55.

Nash, Leonard K. "The Origin of Dalton's Chemical Atomic Theory." *Isis* 47 (1956): 101–16.

Nauenberg, Michael. "Max Planck and the Birth of the Quantum Hypothesis." *American Journal of Physics* 84 (2016): 709–20.

Navau, Carles, Du-Xing Chen, Alvaro Sanchez, and Nuria Del-Valle. "Magnetic Properties of a dc Meta-Material Consisting of Parallel Square Superconducting Thin Plates." *Applied Physics Letters* 94 (2009): 242501.

Newton, Sir Isaac. *Opticks; or, A Treatise of the Reflections, Refractions, Inflections and Colours of Light.* 4th ed. London: William and John Innys, 1730.

Niaz, Mansoor, Stephen Klassen, Barbara McMillan, and Don Metz. "Reconstruction of the History of the Photoelectric Effect and Its Implications for General Physics Textbooks." *Science Education* 94 (2010): 903–31.

O'Brien, Fitz-James. "The Diamond Lens." *Atlantic Monthly*, January 1858, 354–67.

———. "From Hand to Mouth." *New York Picayune*, March 27–May 15, 1858.

———. "The Lost Room." *Harper's New Monthly Magazine,* September 1858, 494–500.

———. "What Was It? A Mystery." *Harper's New Monthly Magazine,* March 1859, 504–9.

———. "The Wondersmith." *Atlantic Monthly,* October 1859, 463–82.

Oersted, Hans Christian. "Experiments on the Effect of a Current of Electricity on the Magnetic Needle." *Annals of Philosophy* 16 (1820): 273–76.

———. "Thermo-Electricity." In *The Edinburgh Encyclopedia,* ed. D. Brewster, 17:715–32. Philadelphia: Joseph Parker, 1832.

Parnell, William J. "Nonlinear Pre-Stress for Cloaking from Antiplane Elastic Waves." *Proceedings of the Royal Society A* 468 (2012): 563–80.

Peacock, George. *Life of Thomas Young, M.D., F.R.S., &c.,* . . . . London: John Murray, 1855.

Pendry, J. B. "Negative Refraction Makes a Perfect Lens." *Physical Review Letters* 85 (2000): 3966–69.

———. "Perfect Cylindrical Lenses." *Optics Express* 11 (2003): 755–60.

Pendry, J. B., A. J. Holden, D. J. Robbins, and W. J. Stewart. "Magnetism from Conductors and Enhanced Nonlinear Phenomena." *IEEE Transactions on Microwave Theory and Techniques* 47 (1999): 2075–84.

Pendry, J. B., A. J. Holden, W. J. Stewart, and I. Youngs. "Extremely Low Frequency Plasmons in Metallic Mesostructures." *Physical Review Letters* 76 (1996): 4773–76.

Pendry, J. B., and S. Anantha Ramakrishna. "Near-Field Lenses in Two Dimensions." *Journal of Physics: Condensed Matter* 14 (2002): 8463.

Pendry, J. B., D. Schurig, and D. R. Smith. "Controlling Electromagnetic Fields." *Science* 312 (2006): 1780–82.

Perrin, J. B. "Discontinuous Structure of Matter." Nobel Lecture, December 11, 1926. The Nobel Prize, www.nobelprize.org/prizes/physics/1926/perrin/lecture/.

———. "Les hypothèses moléculaires." *Revue Scientifique* 15 (1901): 449–61.

Petit, Charles. "Invisibility Uncloaked." *Science News,* November 21, 2009.

Prat-Camps, Jordi, Carles Navau, and Alvaro Sanchez. "A Magnetic Wormhole." *Scientific Reports* 5 (2015): 12488.

Rahm, Marco, Steven A. Cummer, David Schurig, John B. Pendry, and

David R. Smith. "Optical Design of Reflectionless Complex Media by Finite Embedded Coordinate Transformations." *Physical Review Letters* 100 (2008): 063903.

Rayleigh, Lord. "Geometrical Optics." In *Encyclopaedia Britannica,* 1884 ed., 17:798–807.

———. "On Electrical Vibrations and the Constitution of the Atom." *Philosophical Magazine* 11 (1906): 117–23.

Roberts, D. A., M. Rahm, J. B. Pendry, and D. R. Smith. "Transformation-Optical Design of Sharp Waveguide Bends and Corners." *Applied Physics Letters* 93 (2008): 251111.

Rogers, Adam. "Art Fight! The Pinkest Pink versus the Blackest Black." *Wired,* June 22, 2017. www.wired.com/story/vantablack-anish-kapoor-stuart-semple/.

Röntgen, W. C. "On a New Kind of Rays." *Science* 3 (1896): 227–31.

Rutherford, Ernest. "Forty Years of Physics." In *Background to Modern Science,* ed. Joseph Needham and Walter Pagel, 47–74. New York: MacMillan, 1938.

———. "The Scattering of α and β Particles by Matter and the Structure of the Atom." *London, Edinburgh, and Dublin Philosophical Magazine and Journal of Science* 21 (1911): 669–88.

San Francisco Examiner, statement on Ambrose Bierce assignment, January 21, 1896, 6.

Schott, G. A. "The Electromagnetic Field Due to a Uniformly and Rigidly Electrified Sphere in Spinless Accelerated Motion and Its Mechanical Reaction on the Sphere, I." *Proceedings of the Royal Society A,* 156 (1936): 471–86.

———. "The General Motion of a Spinning Uniformly and Rigidly Electrified Sphere, III." *Proceedings of the Royal Society A* 159 (1937): 548–70.

———. "On the Spinless Rectilinear Motion of a Uniformly and Rigidly Electrified Sphere, II." *Proceedings of the Royal Society A* 156 (1936): 487–503.

———. "The Uniform Circular Motion with Invariable Normal Spin of a Rigidly and Uniformly Electrified Sphere, IV." *Proceedings of the Royal Society A* 159 (1937): 570–91.

———. "II. On the Electron Theory of Matter and the Explanation of Fine Spectrum Lines and of Gravitation." *London, Edinburgh, and Dublin Philosophical Magazine and Journal of Science* 12 (1906): 21–29.

———. "V. On the Reflection and Refraction of Light." *Proceedings of the Royal Society of London* 5 (1894): 5526–30.

———. "XXII. On Bohr's Hypothesis of Stationary States of Motion and the Radiation from an Accelerated Electron." *London, Edinburgh, and Dublin Philosophical Magazine and Journal of Science* 36 (1918): 243–61.

———. "LIX. The Electromagnetic Field of a Moving Uniformly and Rigidly Electrified Sphere and Its Radiationless Orbits." *London, Edinburgh, and Dublin Philosophical Magazine and Journal of Science* 15 (1933): 752–61.

———. "LIX. On the Radiation from Moving Systems of Electrons, and on the Spectrum of Canal Rays." *London, Edinburgh, and Dublin Philosophical Magazine and Journal of Science,* 13 (1907): 657–87.

———. "LXXXVIII. Does an Accelerated Electron Necessarily Radiate Energy on the Classical Theory?" *London, Edinburgh, and Dublin Philosophical Magazine and Journal of Science* 42 (1921): 807–8.

Schurig, D., J. J. Mock, B. J. Justice, S. A. Cummer, J. B. Pendry, A. F. Starr, and D. R. Smith. "Metamaterial Electromagnetic Cloak at Microwave Frequencies." *Science* 314 (2006): 977–80.

"A Severe Strain on Credulity." *New York Times,* January 13, 1920.

Shelby, R. A., D. R. Smith, and S. Schultz. "Experimental Verification of a Negative Index of Refraction." *Science* 292 (2001): 77–79.

Silveirinha, Mário G., Andrea Alù, and Nader Engheta. "Parallel-Plate Metamaterials for Cloaking Structures." *Physical Review E* 75 (2007): 036603.

Sinclair, Carla. "Invisibility Cloak Demoed at TED2013." *Boingboing,* February 25, 2013. https://boingboing.net/2013/02/25/invisibility-cloak-demoed-at-t.html.

Smith, David R., and Norman Kroll. "Negative Refractive Index in Left-Handed Materials." *Physical Review Letters* 85 (2000): 2933–36.

Smith, D. R., W. J. Padilla, D. C. Vier, S. C. Nemat-Nasser, and S. Schultz. "Composite Medium with Simultaneously Negative Permeability and Permittivity." *Physical Review Letters* 84 (2000): 4184–87.

Smith, Robert. *A Compleat System of Opticks.* Cambridge: Printed for the author, 1738.

———. *Harmonics; or, The Philosophy of Musical Sounds.* 2nd ed. Cambridge: T. and J. Merrill Booksellers, 1759.

Smith, Thorne. *Skin and Bones.* Garden City, N.Y.: Doubleday Doran, 1933.

Sommerfeld, Arnold. *Optics.* New York: Academic, 1964.

Southey, Robert. *The Doctor.* London: Longman, Brown, Green and Longmans, 1848.

Stark, Johannes. *Prinzipien der Atomdynamik: Die Elektrischen Quanten.* Leipzig: Hirzel, 1910.

Starrett, Vincent. *Ambrose Bierce.* Chicago: Walter M. Hill, 1920.

Stenger, Nicolas, Manfred Wilhelm, and Martin Wegener. "Experiments on Elastic Cloaking in Thin Plates." *Physical Review Letters* 108 (2012): 014301.

Stone, W. Ross. "Nonradiating Sources of Compact Support Do Not Exist: Uniqueness of the Solution to the Inverse Scattering Problem." *Journal of the Optical Society of America* 70 (1980): 1606.

Tachi, S. "Telexistence and Retro-Reflective Projection Technology (RPT)." In *Proceedings of the 5th Virtual Reality International Conference,* ed. S. Richir, P. Richard, and B. Taravel, 69/1–69/9. Angers, France: ISTIA Innovation, 2003.

Thomson, J. J. "XXIV. On the Structure of the Atom: An Investigation of the Stability and Periods of Oscillation of a Number of Corpuscles Arranged at Equal Intervals around the Circumference of a Circle; with Application of the Results to the Theory of Atomic Structure." *London, Edinburgh, and Dublin Philosophical Magazine and Journal of Science* 7 (1904): 237–65.

Thone, Frank. "Cloaks of Invisibility." *Science News-Letter* 45 (1944): 90–92.

Tsakmakidis, K. L., O. Reshef, E. Almpanis, G. P. Zouros, E. Mohammadi, D. Saadat, F. Sohrabi, N. Fahimi-Kashani, D. Etezadi, R. W. Boyd, and H. Altug. "Ultrabroadband 3D Invisibility with Fast-Light Cloaks." *Nature Communications* 10 (2019): 4859.

Valentine, Jason, Jensen Li, Thomas Zentgraf, Guy Bartal, and Xiang Zhang.

"An Optical Cloak Made of Dielectrics." *Nature Materials* 8 (2009): 568–71.

Verne, Jules. *The Secret of Wilhelm Storitz*. Translated and edited by Peter Schulman. Lincoln: University of Nebraska Press, 2011.

Vesalago, Viktor G. "The Electrodynamics of Substances with Simultaneously Negative Values of ϵ and μ." *Soviet Physics Uspekhi* 10 (1968): 509–14.

Ward, A. J., and J. B. Pendry. "Refraction and Geometry in Maxwell's Equations." *Journal of Modern Optics* 43 (1996): 773–93.

Wells, H. G. *Experiment in Autobiography*. Boston: Little, Brown, 1962.

———. *The Invisible Man*. New York: Harper and Brothers, 1897.

———. *Seven Famous Novels*. New York: Alfred A. Knopf, 1934.

Weyl, H. "Über die asymptotische Verteilung der Eigenwerte." *Nachrichten von der Gesellschaft der Wissenschaften zu Göttingen* (1911): 110–17.

Wheaton, Bruce R. "Philipp Lenard and the Photoelectric Effect, 1889–1911." *Historical Studies in the Physical Sciences* 9 (1978): 299–322.

Wiener, Otto. "Stehende Lichtwellen und die Schwingungsrichtung polarisirten Lichtes." *Annalen der Physik* 38 (1890): 203–43.

Wikipedia, s.v. "Inverse Problem." Last modified April 5, 2022, 12:13 (UTC). https://en.wikipedia.org/wiki/Inverse_problem.

Williamson, Jack. "Salvage in Space." *Astounding Stories of Super-Science*, March 1933, 6–21.

Winter, William. *The Poems and Stories of Fitz-James O'Brien*. Boston: James R. Osgood, 1881.

Wolf, Emil. "Optics in Terms of Observable Quantities." *Nuovo Cimento* 12 (1954): 884–88.

———. "Recollections of Max Born." *Optics News* 9 (1983): 10–16.

———. "Three-Dimensional Structure Determination of Semi-Transparent Objects from Holographic Data." *Optics Communications* 1 (1969): 153–56.

Wolf, Emil, and Tarek Habashy. "Invisible Bodies and Uniqueness of the Inverse Scattering Problem." *Journal of Modern Optics* 40 (1993): 785–92.

Wood, B., and J. B. Pendry. "Metamaterials at Zero Frequency." *Journal of Physics: Condensed Matter* 19 (2007): 076208.

Wood, R. W. "The Invisibility of Transparent Objects." *Physical Review* 15 (1902): 123–24.

Wylie, Philip. *The Murderer Invisible.* New York: Farrar and Rinehart, 1931.

Yang, Fan, Zhong Lei Mei, Jin Tian Yu, and Tie Jun Cui. "dc Electric Invisibility Cloak." *Physical Review Letters* 109 (2012): 053902.

Yang, Tao, Huanyang Chen, Xudong Luo, and Hongru Ma. "Superscatterer: Enhancement of Scattering with Complementary Media." *Optics Express* 16 (2008): 18545–50.

Young, Thomas. *A Course of Lectures on Natural Philosophy and the Mechanical Arts.* Vol. 1. London: Joseph Johnson, 1807.

———. *A Reply to the Animadversions of the Edinburgh Reviewers.* London: Savage and Easingwood, 1804.

———. "I. The Bakerian Lecture: Experiments and Calculations relative to Physical Optics." *Philosophical Transactions of the Royal Society of London* 94 (1804): 1–16.

———. "II. The Bakerian Lecture: On the Mechanism of the Eye." *Philosophical Transactions of the Royal Society of London* 91 (1801): 23–88.

———. "II. The Bakerian Lecture: On the Theory of Light and Colours." *Philosophical Transactions of the Royal Society of London* 92 (1802): 12–48.

———. "VII. Outlines of Experiments and Inquiries Respecting Sound and Light." *Philosophical Transactions of the Royal Society of London* 90 (1800): 106–50.

———. "XIV. An Account of Some Cases of the Production of Colours, Not Hitherto Described." *Philosophical Transactions of the Royal Society of London* 92 (1802): 387–97.

———. "XVI. Observations on Vision." *Philosophical Transactions of the Royal Society of London* 83 (1793): 169–81.

Zhang, Baile, Yuan Luo, Xiaogang Liu, and George Barbastathis. "Macroscopic Invisibility Cloak for Visible Light." *Physical Review Letters* 106 (2011): 033901.

Zhang, Shu, Chunguang Xia, and Nicholas Fang. "Broadband Acoustic Cloak for Ultrasound Waves." *Physical Review Letters* 106 (2011): 024301.

"Zola's Eulogy." *St. Louis Post Dispatch,* July 30, 1893, 7.

Epigraph Credits

Chapter 6: excerpt from "The Dunwich Horror," by H. P. Lovecraft, with permission from Lovecraft Holdings.

Chapter 10: excerpt from "The Invisible Robinhood," by Eando Binder, copyright © 1939, 1967 by Otto Binder; first appeared in *Fantastic Adventures,* May 1939; used by permission of Wildside Press and the Virginia Kidd Agency, Inc.

Chapter 13: excerpt from "The Invisible Man Murder Case," by Henry Slesar, with permission from the Slesar Estate.

Chapter 14: excerpt from "For Love," by Algis Budrys, copyright © The Algirdas J. Budrys Trust (1962).

Chapter 15: excerpt from "Cloak of Aesir," by John W. Campbell writing as Don A. Stuart, copyright © 1939, 1967 by John W. Campbell; first appeared in *Astounding Science Fiction,* March 1939; used by permission of Wildside Press and the Virginia Kidd Agency, Inc.

Chapter 16: excerpt from "The Invisible City," by Clark Ashton Smith, with permission from CASiana Enterprises.

Appendix A: excerpt from "Beyond the Spectrum," by Arthur Leo Zagat, copyright © 1934, 1962 by Arthur Leo Zagat; first appeared in *Astounding Stories,* August 1934; used by permission of Wildside Press and the Virginia Kidd Agency, Inc.

Index

The letter "f" following a page number denotes a figure.